KING ALFRED'S COLLEGE
WINCHESTER

To be returned on or before the day marked
below :—

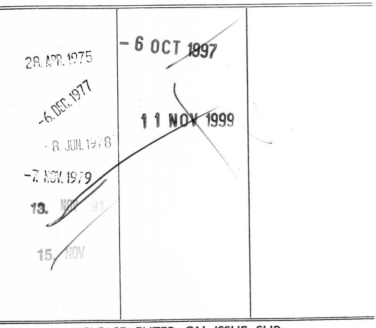

28. APR. 1975

-6. DEC. 1977

- 8. JUN. 1978

-7. NOV. 1979

13. NOV 91

15. NOV

- 6 OCT 1997

1 1 NOV 1999

PLEASE ENTER ON ISSUE SLIP:

AUTHOR ROGERS

TITLE The Common Agricultural policy of Britain.

ACCESSION No. 3 4 3 4 9

AN AGRICULTURAL ADJUSTMENT UNIT SYMPOSIUM

published in conjunction
with The Agricultural Adjustment Unit
University of Newcastle-upon-Tyne

The Common Agricultural Policy and Britain

Edited by
S. T. ROGERS and B. H. DAVEY

SAXON  HOUSE

LEXINGTON BOOKS

Published by

SAXON HOUSE, D. C. Heath Ltd.
Westmead, Farnborough, Hants, England

Jointly with

LEXINGTON BOOKS, D. C. Heath & Co.
Lexington, Mass. U.S.A.

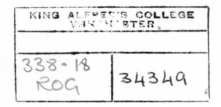

ISBN 0 347 01012 1
LC 72 11269

Printed in Great Britain By Unwin Brothers Limited
The Gresham Press Old Woking, Surrey, England
A member of the Staples Printing Group

Contents

List of Tables

Foreword

During the rest of the 1970s, British agriculture will be faced with the challenge of adapting to the Common Agricultural Policy (CAP) of the European Economic Community (EEC). There will be major changes in the method of supporting farm prices and it can be expected that the marketing channels between the farm gate and the consumer will be modified in various ways. At the same time, entry by Britain into the EEC could have a considerable impact on the pattern of food consumption, on the volume and structure of domestic agricultural production and, in a wider context, on British trade patterns in agricultural products. Moreover, these changes taking place within agriculture will have implications for the rural economy at large.

The papers included in this volume were presented at a conference on 'Agriculture: Britain and the EEC' organised by the Agricultural Adjustment Unit and held in Newcastle-upon-Tyne from 10 to 13 July 1972. This conference was the first major activity of the Unit's programme from 1972 to 1976. Its purpose was to review the various factors bearing on agriculture in the context of an enlarged EEC and the programme was constructed to provide an integrated approach to the problems posed by British membership of the Common Market. The conference was attended by some 120 people, including administrators, academics, extension workers, persons engaged in the agricultural ancillary industries, farmers, landowners and representatives from overseas.

The first three papers have an international flavour. The opening paper discusses the problems of European agriculture in the context of the world agricultural economy and this is followed by papers dealing, respectively, with problems of international trade, especially in temperate-zone products, and the impact of EEC enlargement on the developing 'third' world. There follows a group of five papers that consider more specifically various aspects of enlarging the EEC, particularly so far as agriculture in the United Kingdom is concerned. Recent developments in the CAP are described and their likely effects analysed. The problems and prospects for agricultural institutions, and more specifically agricultural co-operatives, are discussed. Projections of the demand for food, the supply of domestically produced agricultural supplies, farm incomes and structural change are presented. Agricultural trade patterns within the enlarged community are also considered. The last two papers are not, perhaps, so closely related to the theme of the conference, but nevertheless are highly relevant to the problem of agricultural adjustment, which EEC entry will do little to alleviate. The first of these papers discusses the social problems of rural communities, while the second analyses the economic potential of rural areas.

The papers do not represent a statement of the situation as it exists at the present time. They are rather more speculative in character, outlining possible paths of adjustment and adaptation to the new conditions inherent in EEC membership. No attempt was made to reach any conclusions at the conference itself and the papers are presented here without further comment in the hope that they may stimulate thinking and debate generally on the topic of British agriculture and the EEC.

September 1972   S. J. ROGERS
                                                        B. H. DAVEY
                                        Agricultural Adjustment Unit

1 Europe and the World Agricultural Economy

E. M. OJALA,
Assistant Director-General,
Food and Agriculture Organisation of the United Nations.

Introduction

In view of its sponsorship, it can safely be assumed that this conference is concerned essentially with agricultural adjustment. More specifically, its purpose is to examine the ways in which the agricultural sector of the United Kingdom will interact with and adjust to the economic and agricultural systems of the enlarged European Economic Community, when this enlargement becomes a fact.

The first general question to consider is what is expected from agriculture in a development situation. Then there are some important historical elements to discuss which helped to condition the way various European countries treat their agricultural sectors today. It is also necessary to examine the share of Europe in world agricultural output, trade and consumption, and the share of European agriculture as supplier of European demand for food and agricultural products. The evolution of these shares is the main indicator of changes in the relationship between Europe and the world agricultural economy. Since the external impact reflects the degree to which agriculture in European countries is brought into balance with the domestic forces making for change, one must take note of the adjustment problems facing Europe's agriculture, and the policies that governments have been following in relation to them. It is hard to avoid the conclusion that the agricultural policies being followed in Europe have often impeded the internal processes of economic adjustment, and that national maladjustments have been allowed to produce serious and unnecessary difficulties for the world agricultural economy, especially in more recent times. This conclusion raises the question as to whether European countries could solve their domestic agricultural sector problems in ways that would facilitate rather than handicap world agricultural development. Progress towards mutually supporting adjustments in national and world agriculture can hardly be envisaged except within some agreed international framework, and this paper concludes by pointing to some possible elements to be considered if the negotiation of such a framework were to be seriously pursued.

1

Role of agriculture in economic development

It is always useful in approaching questions of agricultural policy to start from first principles. When practice departs seriously from them, it is usually possible to predict the pressures that will be built up, until some degree of reconciliation with the underlying forces making for adjustment is achieved. Although we have become all too accustomed to practical agricultural policies being strongly influenced by short-term expediencies, often for non-economic reasons or because the basic factors were slow to declare themselves, reference to longer-term economic rationality still provides a useful standard, even if only in hindsight.

The development role of agriculture as a sector of the economy is to supply the quantity, quality and variety of agricultural products needed to match the evolution of demand, with steadily rising productivity so as to raise farmers' incomes and to release resources needed for the expansion of non-agricultural growth sectors, which also cater for the demands of consumers. In a mainly closed economy it is the changing pattern of consumption demand on the part of domestic consumers as their incomes rise that is the main determinant of the pattern of production, and of changes in the relative size of the various sectors of the economy. For any given pattern of consumer demand, the relative rates of technological change in the various sectors should also be reflected in the relative sizes of the sectors. As soon as agriculture becomes linked commercially with the rest of the economy, agricultural adjustment is inevitable. The main feature of this adjustment is a relative, but not necessarily absolute, decline in the size of the agricultural sector as economic development proceeds. In the long run, the size and structure of the agricultural sector tend to adjust to the changes in the economy as a whole, but with imperfections and rigidities and at too slow a pace. The common result is that national farming structures, especially in rapidly industrialising countries, tend to be out of date, and farm incomes tend to lag behind incomes in other sectors. This explains the prevalence of governmental intervention in agriculture, which is generally aimed at reducing the human cost of adjustment. These interventions may either facilitate or hinder the basic adjustments that are needed between agriculture and the other sectors of the economy.

In an economy with extensive trading links with the outside world, the intersectoral relationships become much more complex. A country that imports a large proportion of its food requirements or exports a large proportion of its manufacturing or service output may enjoy a high level of both urban and rural prosperity with an extremely small agricultural sector. On the other hand, the consumers in an agricultural exporting country or industrial importing country, including its farming population, may achieve a good standard of living even with an unusually large agricultural sector.

Since most countries are traders, it is clear that there are both national and international dimensions of agricultural adjustment. If agricultural trade were left moderately free, national agricultural adjustments would occur, or be influenced, in the direction of a better balance with both internal and external developments. A few exporters and fewer importers have adopted this approach. For the most part farmers and governments have long rejected the market-place as the sole arbiter of farm prices and incomes, largely because of the well-known rigidities that make it impossible for agriculture to adjust rapidly to the market even under the best conditions. We now live in an era of national policies for agriculture and governments have taken over the responsibility for agricultural adjustment. The prevailing tendency is for countries that can afford to do so, namely the industrialised countries, with relatively small agricultural sectors, to seek adjustments for their agriculture under conditions insulated as far as possible from external influences. Since some 12 to 15 per cent of world agricultural output is traded – many individual countries have a much larger external dependence than this – current national agricultural adjustments in the developed countries have serious distorting effects on adjustments in other countries.

The deep roots of agricultural protectionism

Europe's links with the agriculture of the wider world largely began when the European explorers of the fifteenth and sixteenth centuries returned home with cargoes of spices, indigo, silks, sandalwood, ebony, teak, tea and tobacco. With the onset of the industrial revolution other products of the outer agriculture moved to Europe in swelling flows, especially cotton and the industrial fibres jute, sisal and, much later, rubber. Geographical differences, aided by transport limitations and metropolitan initiatives in colonial development, ensured a high degree of complementarity in the evolution of the domestic and the overseas agricultural systems.

This arrangement experienced its first rude shock when the American colonies, having gained their independence, developed their natural advantages in cereals production and offered increasing quantities of wheat in the European markets in the latter half of the nineteenth century. Essential prerequisites had been the transport revolution in the form of railroads across the prairies and steam-powered steel ships across the Atlantic. Wheat prices in Europe fell drastically, especially after 1880. The farmers' average wheat price in the United Kingdom, which reflected closely the international movements, declined through three decades, from an estimated £13.34 per ton in 1867–9 to £6.08 in 1894–1903.[1]

European countries varied greatly in their responses to this situation. The

3

United Kingdom, which had committed herself to a free trade policy with the repeal of the Corn Laws in 1846 – a policy that supported her leadership as the first great industrial exporting nation – continued resolutely on this course, and became the world's largest importer of food and agricultural products. No measures were taken either to protect agriculture or to facilitate its adjustment to the new situation. Wheat production fell from 2.9 million tons a year in 1867–9 to 1.5 million tons in 1904–10. The area of land under arable farming declined from 18.3 million acres in 1870 to 14.7 million acres in 1910. There was economic distress among arable farmers, but livestock producers were relatively encouraged by the fact that livestock prices did not fall as much as feed grain prices. During the period from 1867–9 to 1904–10, the contribution of livestock products to the gross output of British farms rose from 55 to 75 per cent. This may now be seen as a rational response, though effected by rough reliance on market forces. However, total gross output rose on average by only about 0.25 per cent per annum over this forty-year period. The number of persons occupied in agriculture (about 20 per cent of all occupied persons in 1870), which had already been falling rapidly under the impact of industrialisation, continued to fall after 1880, but more slowly, accounting for about 11 per cent of all occupied persons in 1910.[2] This exodus of workers did not, however, proceed fast enough for farming incomes to maintain their relative position. In 1870 these had been on a par with average incomes in other sectors, but by 1910 they had deteriorated by one-third compared to income developments outside agriculture.

Denmark and the Netherlands were the only other European countries that held firmly to free trade. Denmark had been a grain exporter and responded to overseas competition by converting to livestock production based on cheap imported grains as a deliberate act of adjustment policy. Major efforts were launched, with the aid of farmers' co-operatives, to raise the efficiency and productivity of the livestock industries, and production of milk, butter, eggs and pig meat roughly trebled over the forty years up to 1910. Thus the foundation was laid for a highly prosperous agriculture based on the export of quality livestock products to Britain and Germany. Similar developments took place in the Netherlands. In both these countries, rapidly expanding agricultural sectors made important contributions to the national economy and to the balance of payments.

Elsewhere in Europe, where agricultural trade had just been experiencing a brief interlude of freedom, the response to foreign competition was a gradual return to protection through tariffs on grains and in some cases on livestock products as well. France, Germany and Italy reintroduced protective policies around 1880, and import duties on a wide range of agricultural products, especially wheat, were raised substantially during the 1880s and 1890s as competition from American and Russian grain continued.

Other European countries also introduced protective measures for both agriculture and industry from the late 1870s.[3]

It would take too long to account for the differences in reactions in the various countries. They no doubt resulted however, from the much lower proportion of the population still remaining in agriculture in Britain than on the Continent, the strong feelings against taxes on food in Britain ever since the repeal of the Corn Laws, and the long lead of Britain in the industrial revolution, which meant that redundant farmworkers could readily be absorbed in industries with growing export markets. On the other hand, in France, Germany and generally in Europe, there was a stronger philosophy that rural life should be supported for its own sake and a stronger need for tariffs to protect the earlier stages of industrialisation. Even the United States of America introduced prohibitive tariffs to shelter her growing industries, and in 1890 extended protection to agriculture.

By such measures most European countries protected their arable farming to a great extent from the unfavourable movement of world grain prices. Thus from 1860 to 1900 France and Germany managed to maintain their acreage under wheat and to increase the output of grain products that could have been obtained much more cheaply from overseas. They also seem to have prevented any sizeable exodus of agricultural workers to other sectors, since the percentage of the population in agriculture, which had been in the high 40s around 1860, was still close to 40 at the turn of the centruy. But the livestock industries expanded in most countries. Tracy questions the advantages to the European peasants, as distinct from the large grain farmers, of the protection measures of this period.[4] Certainly the continental farmers came to rely on the tariff for economic support to their industry.

The economic crisis of the 1930s at first strengthened this reliance. Almost everywhere in continental Europe tariffs on farm products were raised, often to levels two to three times the world price. When this failed to protect farm prices and incomes sufficiently, a wide range of non-tariff measures for supporting domestic agriculture made their appearance. In addition, self-sufficiency became a conscious goal of agricultural policy in many European countries, to save foreign exchange as well as to support the concept of retaining people in farming for social, political and even military reasons.[5] Governments tried to shelter their economies and their agriculture from the problems of the time on a purely nationalistic basis, everyone striving to save imports and promote exports. Thus the international economic crisis was deepened and prolonged. By 1931 the flood of subsidized agricultural exports to the United Kingdom was such that the era of free trade in Britain was brought to an end. The main effect, however, following the Ottawa Conference, was to switch import demand from foreign to imperial suppliers, while giving first preference to domestic farmers. Denmark and the Netherlands, too, were forced by the protective policies of their neighbours to intervene in the markets to assist their farmers.

The Second World War and the ensuing balance of payments problems convinced most European countries of the need to maintain at least a minimum level of domestic agricultural production. After the war most countries, including Britain, intensified policies for increasing output, and price supports to farmers were more widely introduced to promote this goal. From this time, agricultural protection in Europe, by insulating farmers increasingly from world prices, degenerated into protectionism. Britain, nevertheless, while protecting her farmers, continued to do so in a manner that permitted her population to benefit from relatively low food prices, reflecting a large inflow of imported agricultural products.

Colonies become nations

Throughout the eighteenth, nineteenth and early twentieth centuries the industrialised countries of Europe invested heavily in their associated territories overseas. These investments promoted the production for export of agricultural products and minerals needed to supplement the supplies of primary products available in Europe itself, and opened up overseas markets for the industrialising countries. There was considerable British investment in countries outside the empire, particularly in South America. Thus, for several centuries a large part of the agricultural export development in the southern hemisphere was designed from Europe to complement European agriculture. For the continental industrial countries of Europe, which continued to protect their own agriculture after 1880, the complementarity was largely that between temperate European and tropical overseas agriculture, any needed imports of temperate-zone products being obtained mainly from the agricultural countries of Eastern Europe and from Russia. For Britain, however, the agricultural complementarity increasingly reflected in her import trade was economic as well as geographical, reflecting her free-trade policy and her investments in temperate countries both within and outside the empire, as well as in her tropical colonies. Her agricultural trade network became, therefore, not only very much larger, but also much more worldwide than that of the continental industrialised countries.

By 1928 the countries of Europe (excluding the USSR) were obtaining 17 per cent of their total imports from overseas dominions or colonies associated with one or another of them. These imports flowed mainly to the United Kingdom (41 per cent), Germany (19 per cent) and France (18 per cent). Europe's imports from the rest of the world (outside Europe) in the same year constituted 29 per cent of total imports. Most of the 17 per cent, and probably at least half of the 29 per cent, were agricultural products.[6]

6

Since the agricultural exports of the colonies, dominions and other southern hemisphere countries were their main exports, and the European countries their main markets, the dependence commercial agriculture in a large of part of the world outside Europe on developments in the European market was very great. As a result of the attempts by European countries in the 1930s to solve their balance of payments problems by a new surge of protection accompanied by stronger imperial preferences, imports from their associated overseas countries and territories rose from 17 per cent of all imports in 1928 to 23 per cent in 1938 ,at the expense of the rest of the world. The proportion of all imports from Europe's overseas empires taken by the United Kingdom rose to 54 per cent. Table 1.1 shows the share of Europe (including the USSR) in world production, consumption and imports of

Table 1.1

Share of Europe (Including the USSR) in World Production, Consumption and Imports of Selected Agricultural Commodities in 1934–8

Commodity	Percentage		
	Production	Consumption	Imports
Wheat	49.0	54.5	85.6
Barley	47.6	50.2	81.5
Maize	19.2	25.6	95.7
Beef	28.5	31.3	
Sugar	30.9	39.8	31.0
Tobacco	17.2	27.0	98.0
Cottonseed	11.2	15.9	98.0
Groundnuts	0.3	28.7	99.0
Copra (and oil)	0	55.7	63.0
Coffee	0	29.8	46.0
Tea	0.6	30.0	68.3
Cocoa	0	53.4	58.8
Cotton	10.7	39.8	91.0
Jute	0	41.5	84.0
Sisal	0	53.5	53.0
Rubber	0	36.0	41.0

Source: *Agricultural Commodities and Raw Materials: Production and Consumption in the Different Parts of the World, 1934–8* (Rome, International Institute of Agriculture, 1944).

selected agricultural products in 1934–8. The striking feature is the predominance of Europe as the market for the world's agricultural exporters (including European), the United Kingdom being by far the largest importer. The table underlines the vulnerability of the world's agricultural economy to factors that might affect the agricultural import requirements of Europe. This vulnerability remains.

The rise of agricultural protectionism in Europe has been accompanied by the winning of political independence on the part of the former colonial territories. Both of these developments have gained momentum since 1945. It is not surprising that the coincidence of these changes has led to the progressive build-up of international pressure for a wholly new look at the trading relationships between the world's exporters and importers of primary products, and in particular at the agricultural policies of the industrialised countries. One of the major recent outcomes of this pressure was the adoption by the United Nations General Assembly in 1970 of an International Development Strategy for the 1970s. It presents, for the first time in history, an officially accepted normative approach to world development. It is relevant to this conference because it includes goals of agricultural output growth (4 per cent per annum) and export growth (somewhat above 7 per cent per annum for all exports) for developing countries that are unattainable without international adjustments in agriculture. Governments have resolved 'to promote a rational system of international division of labour' and to take action 'with a view to ensuring that developing countries have improved access to world markets and to market growth for products in which they are presently or potentially competitive'.

It is a far cry from the adoption of a resolution in the United Nations to the opening up of access to the agricultural markets of the industrialised countries, but it is historic that such a declaration should be formulated, let alone adopted, even with serious reservations on the part of the economically powerful countries directly affected. Its adoption is a recognition by the developed countries that they have continuing and evolving international obligations to facilitate and promote the development of the 'outer' regions of the world, no longer as colonies or spheres of influence, but as interdependent elements in the development of the world economy.

Changes in the role of Europe

The position of non-European countries as suppliers of temperate-zone agricultural products to Western Europe continues to deteriorate. This is apparent from Table 1.2, which presents data on Western Europe's imports of major cereals, meat, butter and sugar during the 1960s, by main areas of origin. The total value of imports of the countries in this region declined between 1963–5 and 1969 for butter and sugar, changed little for barley and maize, increased slightly for wheat and greatly for live animals and meat. However, the proportion of imports originating in Western Europe itself rose markedly for cereals and sugar and less markedly for butter and meat. The proportion of wheat and meat coming from Eastern European countries also increased. The role of the rest of the world as agricultural suppliers to Western Europe declined for all these products, most strongly for wheat and barley. This effect is summarised for a larger number of commodities in Table 1.3. It would appear that only for fresh fruit and vegetables were outside suppliers able to hold their place during this period.

These developments are looked at from a broader point of view in Table 1.4,

Table 1.2

Western Europe's Imports of Wheat, Barley, Maize, Meat, Butter and Sugar by Areas of Origin, 1963–5 to 1969

Product and year	Total	Origin – percentage of total		
	$ million US	Western Europe	Eastern Europe	Rest of world
Wheat 1963–5	859.1	16.3	2.1	81.6
1966–8	840.1	27.0	4.9	68.1
1969	1,031.5	45.5	8.1	46.4
Barley 1963–5	291.4	54.8	5.2	40.0
1966–8	348.6	73.1	2.6	24.3
1969	321.0	79.9	3.6	16.5
Maize 1963–5	970.4	11.1	5.4	83.5
1966–8	1,164.3	16.1	2.0	81.9
1969	1,051.3	22.1	1.9	77.0
Meat 1963–5	2,653.4	58.6	9.0	32.4
1966–8	3,103.5	59.9	12.8	27.3
1969	3,687.3	62.1	12.9	25.0
Butter 1963–5	496.5	43.5	5.1	51.4
1966–8	461.4	52.1	7.0	40.9
1969	427.3	55.4	2.9	41.7
Sugar 1963–5	864.7	16.8	10.7	72.5
1966–8	578.7	16.0	7.5	76.5
1969	599.9	26.1	6.0	67.9

Source: Economic Commission for Europe: *Agricultural Trade in Europe* (United Nations, New York, 1971).

9

which compares matrices of Europe's agricultural trade among groupings of European countries and with the rest of the world in 1960–2 and in 1967–9. During this seven-year period the value of the agricultural imports of European countries from all sources increased by about one-third, but as a share of world agricultural trade it fell slightly, from 59 to 57 per cent. The table demonstrates that the dependence of the world's agricultural exporters on Europe, and particularly on Western Europe, is still remarkably high. But the most striking feature of the table is the development of intra-trade among the various European groupings. Over the period of comparison, world agricultural imports increased by 38 per cent while the imports of Western Europe from sources outside Europe increased by only 12 per cent. Among Western European countries agricultural trade expanded by 73 per cent, and among members of the EEC by no less than 143 per cent.

Since 1967, when the EEC Common Agricultural Policy began to have an effect, the above trends no doubt include the influence of the progressive implementation of the policy. However, the same trends towards agricultural self-sufficiency are evident elsewhere in Western Europe outside the Six. The main common factor is the strengthening of agricultural protection everywhere in Western Europe; the CAP seems to be the most integrated and powerful manifestation of a general determination on the part of industrial-

Table 1.3

Imports from Countries Outside Western Europe as a Proportion of the Area's Total Gross Imports*

| | Percentages | | | |
Product	1963–5	1966–8	1968	1969
Wheat	83.7	73.0	63.8	54.5
Barley	45.2	28.1	19.0	20.1
Maize	88.9	83.9	84.4	78.9
Meat and live animals	41.4	40.1	35.9	37.9
Butter	56.5	47.9	42.7	44.6
Cheese	24.9	18.5	16.7	10.7
Eggs	32.6	33.2	27.3	21.8
Fresh fruit	53.4	52.9	52.9	50.5
Fresh vegetables	32.1	33.3	32.4	33.1
Sugar and honey	83.2	84.0	80.5	73.9

Source: *United Nations, Commodity Trade Statistics.*
* Calculated from value data.

Matrices of Europe's Agricultural Trade, 1960–2 to 1967–9.

US $ million

From \ To	Europe	Western Europe	EEC	Other W. Europe	Eastern Europe	USSR	Rest of World	World
Europe	10,362	8,039	4,440	3,599	1,861	462	2,881	13,243
Western Europe	7,408	6,960	3,947	3,013	340	108	2,469	9,877
EEC	3,280	3,152	2,094	1,058	100	28	1,160	4,440
Other	4,128	3,808	1,853	1,955	240	80	1,309	5,437
Eastern Europe	1,554	658	339	319	542	354	140	1,694
USSR	1,400	421	154	267	979	–	272	1,672
Rest of world	12,954	11,295	5,890	5,405	761	898	13,273	26,227
World	23,316	19,334	10,330	9,004	2,622	1,360	16,154	39,470

annual average 1967–9

From \ To	Europe	Western Europe	EEC	Other W. Europe	Eastern Europe	USSR	Rest of World	World
Europe	16,603	13,781	8,576	5,205	2,119	703	4,497	21,100
Western Europe	12,749	12,064	7,624	4,440	505	180	3,414	16,163
EEC	6,970	6,760	5,096	1,664	181	29	1,524	8,494
Other	5,779	5,304	2,528	2,776	324	151	1,890	7,669
Eastern Europe	2,365	1,216	722	494	626	523	322	2,687
USSR	1,489	501	230	271	988	–	761	2,250
Rest of world	14,719	12,676	7,027	5,649	898	1,145	18,764	33,483
World	31,322	26,457	15,603	10,854	3,017	1,848	23,261	54,583

percentage increase 1960–2 to 1967–9

From \ To	Europe	Western Europe	EEC	Other W. Europe	Eastern Europe	USSR	Rest of World	World
Europe	60	71	93	45	14	52	56	59
Western Europe	72	73	93	47	49	67	38	64
EEC	113	115	143	57	81	4	31	91
Other	40	39	36	42	35	89	44	41
Eastern Europe	52	85	113	55	15	48	130	59
USSR	6	19	49	1	1	...	180	35
Rest of world	14	12	19	5	18	28	42	28
World	34	37	51	21	15	36	44	38

Source: *UNCTAD Handbook of International Trade and Development Statistics.*
Notes: *Agricultural Trade* covers *SITC* 0, 1, 2 (less 27 and 28) and 4 (i.e. includes fishery products and crude forestry products.

Valuation millions of *US* dollars f.o.b. based on export figures.

ised countries to preserve more of their agricultural markets for their own farmers, at a high level of cost to consumers and taxpayers. Thus over the period 1955–7 to 1967–9 self-sufficiency in wheat increased in the EEC from 88.7 to 107.3 per cent, and in the four candidate countries [7] from 37.6 to 47.2 per cent. Self-sufficiency in sugar has fluctuated in the EEC but was 102.4 per cent in 1967–9 compared with 96.0 per cent in 1955–7.[8] In the Four, sugar self-sufficiency was over 37.5 per cent in 1964–9 compared with 31.7 per cent in 1958–60 and 37.3 per cent in 1955–7.

It appears that the Four have already gone far in preparation for membership of the EEC in shifting certain agricultural imports from the rest of the world to one another and to the Six. This is evident from Table 1.5, which shows matrices of the imports of the EEC and the candidate countries for wheat, barley, maize, meat, vegetable oilseeds and oils and sugar for the two periods 1959–61 and 1967–9. It shows that between these two periods, the EEC members greatly increased their imports of:

(a) wheat, from within the EEC, at the expense of the rest of the world;
(b) barley, from within the EEC but also from the candidate countries at the expense of the rest of the world;
(c) maize, from within the EEC, from Eastern Europe and, in particular, from the rest of the world;
(d) meat, from all sources;
(e) vegetable oilseeds, from within the EEC and from Eastern Europe;
(f) sugar, from within the EEC and from Eastern Europe.

What is equally interesting is that over the first eight years of the 1960s the four candidate countries purchased much more wheat from the EEC at the expense of extra-European suppliers, more barley from each other but much less from outside Europe, and much more maize from the EEC, while maintaining purchases from the outside.

There are other studies that show that similar trends were occurring well before the EEC was established. Thus over the period 1951–9 North America and Asia were already being displaced by Eastern Europe, EEC members, other non-sterling European countries and South and Central America as suppliers of food and agricultural commodities to the EEC member countries.[9]

The developing countries, despite their capacity to supply tropical food and beverage products that do not compete with European agriculture, seem to have fared almost as badly as the USA in the low rate of their participation in the expansion of EEC imports of food products and livestock between 1963 and 1969.[10] A factor in this may be the development in Europe of synthetic products and substitutes for agricultural raw materials exported by developing countries, especially cotton, jute and hard fibres. In general,

Table 1.5

Matrices of Imports of the EEC and the Candidate Member Countries[1] for Selected Agricultural Commodities, 1959-61 and 1967-69

From / Imports of	EEC 1959-61	EEC 1967-9	Candidate members 1959-61	Candidate members 1967-9	USSR and Eastern Europe 1959-61	USSR and Eastern Europe 1967-9	Rest of the world 1959-61	Rest of the world 1967-9	Total 1959-61	Total 1967-9
					Thousand metric tons					
Wheat[2]										
EEC	757	1,815	28	1	555	295	3,749	3,439	5,089	5,550
Candidate members	289	1,291	64	15	396	644	4,080	2,830	4,802	4,780
Barley										
EEC	507	1,725	269	484	175	54	1,441	873	2,392	3,136
Candidate members	124	142	50	119	178	–	1,043	383	1,395	644
Maize										
EEC	164	1,019	–	1	198	284	3,508	8,947	3,870	10,251
Candidate members	295	965	–	–	112	93	3,024	2,911	3,431	3,969
Meat[3]										
EEC	201	621	98	134	42	141	212	432	553	1,328
Candidate members	12	13	55	157	4	9	792	649	863	828
Vegetable oils[4]										
EEC	70	271	18	5	28	241	910	919	1,026	1,436
Candidate members	20	58	3	6	1	55	372	388	396	507
Sugar[5]										
EEC	180	482	30	45	49	275	949	962	1,208	1,764
Candidate members	92	102	93	103	211	170	2,731	3,188	3,127	3,017

[1] United Kingdom, Ireland, Denmark and Norway
[2] Excluding wheat flour
[3] Meat, fresh chilled or frozen
[4] Oilseeds, oil nuts and oil kernels
[5] Sugar and honey

Source: Foreign Trade Series C. (OECD, Paris).

it would appear that the long-standing preferences and special trading relationships between the industrialised European countries and their former colonies, in both tropical and temperate zones, are being weakened by both the preoccupations of the former with the exigencies of their own agricultural integration and the pressures of other developing countries for more generalised preferences in favour of poorer nations. There is scope for expansion in consumption and therefore in the imports of tropical products, especially beverages and bananas, on the part of Eastern European countries, where consumption is artificially limited by trade restrictions.

All those concerned with the further expansion of world agricultural trade as the instrument for the more rapid and balanced development of all countries, especially those that have no economic alternative to agricultural exports at this stage of their history, could not but view with alarm any prospect of the United Kingdom and Denmark surrendering the last remnants of their open agricultural trading traditions in the course of their integration into the EEC.

In 1971, FAO completed a new round of world projections of demand and production of agricultural commodities over the period 1970–80. This work was done before the negotiations for the enlargement of the EEC were finalised. But, in a parallel exercise, an attempt was made to assess the effect on the commodity projections for the four candidate countries to 1980 if they adhered to the CAP and adjusted their prices to the 1969/70 EEC level over a transition period of four and a half years beginning on 1 January 1973, and maintained that level up to 1980.[11] The study was not of course a forecast. It set out to examine how demand and production might change in the Four, on the unlikely assumption that EEC agricultural policies and prices remained the same in 1980 as in 1970. The results pointed to some expansion in production of, and some dampening in the demand for, most agricultural commodities in aggregate in the four applicant countries, as compared with the standard projections at constant 1970 prices. The effect of these changes on the earnings potential of exporters outside the enlarged Community could be quite substantial. The impact of the assumed price changes in the Four on the projection of the net import balance of Western Europe in 1980 for main temperate-zone products was a reduction of about one-third, from 3.4 to 2.2 billion US dollars. This was not an economic analysis of the enlargement of the EEC, but simply a study of alternative price effects on demand and production, and hence on the net trade balance. Because of the influence of agricultural protection in Europe generally, similar trends could appear without reference to the EEC.

Is it inevitable that agricultural protection in industrialised countries must run its course? Is the mature national or regional economy to be one in which population and living levels will be adjusted to the capacity of the land to provide food up to, but not beyond, self-sufficiency for the nation or group, at whatever cost? Must we accept that it is beyond the wit and will of man, who is increasingly taking control of his economic life, to utilise the agricultural resources of the world to the better advantage of all, through the exchange of goods?

Agricultural adjustment

We have to learn from the lessons of the past, and give more weight in agricultural policy-making to facilitating and promoting the integration of agriculture with the evolution of the economy as a whole. It is probably true to say that in most of continental Europe agriculture emerged from the industrial transformation up to the Second World War still relatively insulated not only from external factors, but also from the internal modernisation of the European economies.

During the inter-war period agriculture in Europe showed many symptoms of a depressed industry – antiquated structure, low wages, low profits, slow responses in output and productivity and a wasteful use of manpower. In 1934–8 crop yields in many countries were not much higher than in 1909–13. The war stimulated the application of science to agriculture. Since the war the pressures of economic and social conditions, as well as continued advances in biology and technology, have been demanding adaptations in European farming on a historically unprecedented scale. After a century of being relatively sheltered from the necessity for change, the adjustment of agriculture in Europe has been lagging behind the exigencies of the time. The difficulties of agricultural policy have accumulated over the years and are now immense, in political, economic, and human and social terms.

The factors to which European farming has been slow to adjust in this century have been largely internal, as distinct from the previous century when the fundamental challenge came from the outside. The major challenge to traditional structures in the post-war years has been the impact of technology. The rate of increase in crop yields, in the productivity of farm animals, in mechanisation and in volume of output suddenly accelerated in the 1950s and 1960s. Improved varieties of crops, new and improved feeding methods for livestock and new chemicals for enhancing soil fertility and controlling diseases and pests have been introduced and widely adopted. Further innovations are on the way. Farming is becoming increasingly linked with other sectors of the economy, purchasing an increasing proportion of its inputs from the chemical and feed-mixing industries and selling products for further processing to the food industries. European agriculture has acquired the capacity to expand production at a speed never before possible, but has become more vulnerable than before to commercial fluctuations.

15

This upsurge in productivity has hastened the delayed evolution of a relative decline in the size of the agricultural sector. In Europe this takes the form of an absolute decline in the size of the farm labour force. Over the last ten to fifteen years there has been a net outflow of labour from agriculture in European countries of 3 to 6 per cent a year. But for decades average farm incomes have been well below average non-farm incomes and the labour outflow has not been fast enough to remedy this gap. A modern balanced industrial economy needs only 5 to 8 per cent of its labour force in agriculture, as in for example the USA, Belgium and the Netherlands; in 1965–7 France still had 16.2 per cent and Italy 24.4 per cent in this sector.

The relative improvement of farm incomes seems to have taken precedence over structural modernisation as the main objective of agricultural policies. Western Europe has traditionally had a small-farm structure. In response to the rising productivity of the factors of production, the average size of farms has been increasing and their total number falling as a result of amalgamations and extensions. The annual rate of reduction in numbers of farms during the 1960s has varied from over 4 per cent in the Netherlands and Sweden and 3.7 per cent in Belgium, to 2.9 per cent in France and 0.7 per cent in Italy. Most authorities believe that this adjustment of the organisation of farms to modern technological requirements is proceeding far too slowly and that only a small minority of the farms in Western Europe are large enough to exploit fully the present and future opportunities. The vast majority are still too small to do so, and hence provide unacceptably low incomes to their operators. Here is the core of the farm income problem. There is a duality in the agricultural sectors of advanced Western European countries, namely a 'production' sector in which a small proportion of the total farm labour force produces the bulk of the commercial output and a 'small-farm sector' consisting of a large number of farmers whose enterprises are too small ever to be viable, and some part-time farmers who derive much of their income from non-farm sources.

It is unfortunate for European agriculture – and for the rest of the world – that the tremendous rise in European farm productivity has coincided with the slowing down of demand for basic foods in European countries. This stabilisation of demand is due to slower population increases and to the saturation of *per caput* demand at the prevailing high income levels, except for meat and fruit. While the rate of growth of aggregate food demand in Western Europe has been falling from over 3 per cent per annum in the 1950s to about 1.6 per cent per annum in the 1970s, the rate of increase in agricultural output has risen from 0.5 per cent per annum in the 1930s and 1940s to 3 per cent per annum in the 1950s and 1960s. Western Europe has now come to share with the USA the built-in tendency to produce a domestic surplus of nearly all temperate-zone agricultural products except beef. The

16

possibility of using this capacity to supply the import requirements of developing countries has been greatly reduced in recent years by the technological breakthrough in tropical cereals production, especially in the formerly large food deficit countries of Asia. This means that the further modernisation of Europe's agriculture, for which there is still great scope, must be accomplished with only a slow rate of expansion in total output.

In this combination of circumstances, the policy of supported prices for farmers adopted by virtually all Western European governments has had mixed results. High price supports have certainly stimulated output and modernisation and contributed to the imbalance of supply and demand in domestic markets. Usually, in the absence of production controls and consistent stockholding policies they have led to surpluses, a large part of which have had to be exported haphazardly at subsidised prices. Price supports have stabilised and improved the incomes of farmers, especially the large ones, and retained in farming many resources of land, labour and capital, especially in small farms, which technically are not needed to produce the desired output. However, they have not solved, and cannot solve, the basic problem of the small non-viable farms. This must be recognised and treated as a social problem. Measures taken at the border, that is import controls and export subsidies, are essential to perpetuate the high-support-price approach that dominates continental European policies. With the addition of variable import levies the protection package becomes ever more costly to European taxpayers and consumers as agricultural productivity improves elsewhere, without offering any hope or prospect of building a rational and economic agricultural sector, well adapted to the needs of the economy as a whole.

When a large number of countries, as in Europe, wealthy enough to protect and subsidise their agriculture, either at home or in export markets or both, all pursue the national policies outlined above, the external impact can become the dominant factor in world markets. This is the current situation for most temperate-zone commodities. If such policies result in world-market imbalance for a commodity the rigidities of protectionism force the adjustment upon the exporter, who is least able to compete in the provision of export aids, and who is usually a developing country or a country heavily dependent on agricultural exports, however low-cost their production. The location of production adjustments as imposed by the forces now so highly organised in Europe is bound to lead to economic losses for the world as a whole.

Is international agricultural adjustment possible?

The concern here is not specifically with whether the 1973 GATT negotiations (including agricultural barriers) or the UNCTAD confrontations will be successful, or whether more commodity agreements will be concluded. It is the more basic question of whether a convergence of the long-run objectives of national and international agricultural adjustment cannot soon be identified, making it possible to envisage an internationally negotiable package of goals and actions designed first to stabilise the levels and costs of agricultural protection and then to lead over a longer term to their balanced reduction, thereby permitting a wider dispersion of agricultural production and trade expansion to the advantage of all.

Clearly it is not simply a question of trade barriers. It is equally a question of national agricultural policies, and their social, economic and political setting. In nearly all developed countries the State has largely taken over from the brute forces of the market-place the responsibility for adjusting agriculture to the changing world and it is hard to envisage anything like a return to free trade in agricultural products – freer trade perhaps, but internationally managed. All important trading countries have forged sophisticated instruments of national policy for agriculture. These could be used to implement a more balanced mixture of desired national and international adjustment goals, if the latter carried any weight. It is a thesis of this paper that the national goals of improving farm incomes and farm structures in developed countries are compatible over the longer term with the international goal of expanding agricultural trade.

In a recent study, FAO[12] has distinguished agricultural adjustment problems facing the developed countries at three levels – the international market, the national market and the farm. The maladjustments at all three levels are related. They combine to hold back the full application of science and technology to agriculture both in Europe and abroad, and impede progress in raising the living standards of farm people and agricultural countries in many parts of the world. At the international level, where the trade and development problems of other countries are affected, the developed countries have difficulty in committing themselves to any reduction of subsidised exports or to the opening up of access because of the pressure of domestic surpluses on their national markets. Their national market imbalances result from their virtual failure so far to find a way of moving enough resources out of agriculture in an orderly manner and rationalising individual farm structures rapidly enough. Attempts to deal with the international problems may have very limited success unless simultaneous attempts are made to deal with the other two.

Much more analytical work is needed to diagnose the problems at farm and national market levels in developed countries so that suggestions for dealing with them more effectively can be formulated to enable their farm sectors to become economically viable without requiring such high levels of protection. At the level of improving farm structure many developed countries have adopted policies which are having success to the extent of their coverage.[13] The efforts of governments to strengthen greatly such policies and speed up their implementation are highly relevant. Persistent long-term measures are needed to facilitate and humanise the exodus of farmers and workers who wish to leave non-viable holdings and to ensure, on a national scale, the aggregation of farm land into viable commercial enterprises. A key to success will be the separation for policy purposes of income support for farmers on non-viable holdings, which is a social problem, from the formulation of price policies designed to orientate commercial production.

The concept of a 'package' for international negotiation would be essential. The package would have to reflect the existence of a number of interlinked objectives of industrialised and developing countries, and of a diversity of national conditions and instruments. The agricultural adjustment goals of developed countries would be made an integral part of the approach, linked with the expansion of import demand facing the agricultural export sector of developing countries. Would it be possible for all the European and other developed countries concerned to negotiate such an agreed package of inter-related quantitative objectives at all three levels, to be achieved by the most appropriate national methods, with suitable provision for international consultations on progress at intervals of a year or two? If so, the international consequences might add up to a progessive measured expansion of multilateral trade, of which a share to be negotiated could be reserved for developing countries. FAO is studying these concepts intensively, in preparation for presenting an analysis of the issues of international agricultural adjustment to the FAO Conference in 1973. Such a broad framework, if it could be agreed, would provide a reference for trade negotiations in any appropriate forum, either on a broad front or on a particular commodity.

There are no doubt other approaches that deserve serious consideration. Whatever the detailed measures that may be thought appropriate, one thing is clear. The alternative to some internationally agreed move to resolve agricultural adjustment problems in European and other industrialised countries will be continuing social problems in their rural sectors, rising economic and political tensions with the rest of the world and the distortion of agricultural and economic growth in many developing countries that are already extremely poor. It would be greatly in the interest of the world agricultural economy if the enlargement of the EEC were to permit more weight to be given in the formulation and implementation of agricultural policy in Europe to international adjustment problems, especially in so far as they affect the trade and development of the developing countries.

References

1) Ojala, E. M., *Agriculture and Economic Progress* (OUP, London, 1952), p. 193.
2) Ojala, E. M., *op. cit.*, p. 85.
3) Tracy, Michael, *Agriculture in Western Europe: Crisis and Adaptation since 1880* (Jonathan Cape, London, 1964).
4) Tracy, Michael, *op. cit.*
5) *World Agriculture: An International Survey*, Royal Institute of International Affairs (OUP, London, 1932), Ch. VIII.
6) Statistics taken from *Europe's Trade* (League of Nations, Geneva, 1941).
7) This paper was finalised before the result of the Norwegian referendum was known.
8) Calculations based on official EEC data.
9) Schmidt, S. C., *Commodity Structure and Regional Distribution of EEC Imports: The Formative Years 1951–1959. Part 1 – Food, Beverages and Tobacco and Oils and Fats*, University of Illinois College of Agriculture, Research Report AERR-70 (February 1965).
10) See *A Future for European Agriculture. A Report by a Panel of Experts*, Atlantic Papers 4 (Atlantic Institute, Paris, 1970), Table 11.
11) *Implications of the Possible Enlargement of the EEC for Agricultural Commodity Projections, 1970–80*, Projections Research Working Paper No. 6 (FAO, Rome, 1971).
12) *Agricultural Adjustment in Developed Countries*, Document ERC 72/3 (FAO, Rome, June 1972).
13) See Butterwick, M. and E. N. Rolfe, 'Structural Reform in French Agriculture – the Work of the SAFERS', *JAE*, *XVI*, no. 4 (1965);
Whitby, M. C., 'Lessons from Swedish Farm Structure Policy', *JAE*, *XIX*, no. 3 (1968); and *Structural Reform Measures in Agriculture* (OECD, Paris, 1972).

2 Problems of International Agricultural Trade

J. A. SCHNITTKER,
Schnittker Associates,
Washington, DC.

There are so many problems but no established hierarchy to adhere to in discussing 'Problems of International Agricultural Trade'. They have been discussed and written about endlessly and they have been brooded over in conferences like this. One is hard pressed to say anthing new, and to go beyond the conventional slogan that advises that 'we must negotiate improvements in domestic policies in order to expand agricultural trade'.

There are times when it seems that we take the apparent dangers to trade too seriously, for the circumstances causing expansion of international trade in agricultural products are so overwhelming that one need not be terribly concerned about the myriad of small and detailed operational problems that governments and traders face. Solutions, or at least conclusions, to old problems need to be sought, of course, but the world does not end when they are not found. There are new and expanding markets to be explored in Eastern Europe, the Soviet Union and China, for example. Addressing ourselves to these matters can serve to divert the USA, Canada and Australia from dwelling for ever on the perversity of the variable levy system of the Common Agricultural Policy, or on the amount of grain we could export to Europe if prices guaranteed to European farmers were not so high.

At other times it is easy to become pessimistic about the attempts to improve the prospects for increasing international trade. Trade talks have a tendency to try to solve the problems of the past. This tendency is not limited to international affairs, and certainly not to agricultural affairs. It is observed in nearly all the countries of the world in connection with domestic agricultural policies. We are certainly addicted to it in the USA, where political debate often focuses on 'saving the small family farmer' or 'raising farm prices'. The results of political action, however, seem only to speed the disappearance of the small farmer, and the ·imperatives of trade expansion severely limit the possibilities for raising prices. This tendency to dwell on the past asserts itself especially in actions by governments in attempting to recover trading advantages that were set aside in the 1960–1 round of worldwide tariff negotiations, in the UK cereals arrangement of 1964, or in the Kennedy Round in the late 1960s. Perhaps there is some value in this exercise for it fills negotiators' time when it might otherwise be wasted while waiting for opportunities to negotiate current questions. Possibly the rituals of economic diplomacy require unrelenting efforts to salvage what error or circumstances have lost or set aside, as much as due diligence to minimise or avoid those errors or dangers hovering in our future.

It seems rather useless to negotiate over past trivia and better to take today as it is and work to establish both a method and a timetable for the future, based upon a realistic assessment of the factors that have limited, and are limiting, improvements in the accepted rules of international agricultural trade.

Agricultural trade problems in context

The questions of world agricultural adjustment, of incomes of farm operators and agricultural workers relative to workers in other economic sectors and the prospects for expansion of agricultural trade continue to be of broad interest, not only to agricultural specialists, but to people in political life and to industrial and monetary specialists. For several years there has been growing zeal and knowledge on the part of American businessmen, whose daily bread is earned in steel, shoes, textiles or chemicals and who can discuss the intricacies and inadequacies of the CAP or the US Agricultural Act of 1970, or the benign influence of soya beans on the balance of payments. The adverse effects of agricultural policies and programmes in nearly all countries on the expansion of trade are widely appreciated, especially outside the country whose policy is under discussion, and the importance of doing better in the future is generally agreed. This is not terribly surprising. It is not easy to overstate the extent of confusion and even anarchy in world agricultural trade in recent years. Liberalisation of agricultural trade rules was attempted in the Dillon Round in about 1960 and in the later Kennedy Round of trade negotiations, with the most limited results. Yet agricultural trade has expanded, although not for every country or in every commodity. For the future, the various public or private projections that have been made point to a sharp increase in agricultural trade over the next ten to twenty years, especially in grains and oilseeds.

Whether or not these past efforts at accelerated trade expansion were realistic is certainly open to question. It has been stated on another occasion that in the Kennedy Round '...the United States had no discernible master strategy in regard to winning support at home for negotiating domestic policies', partly because United States representatives never considered this a realistic possibility. Also, '...it was clear to many Americans involved in the negotiation that solidifying the European Community at the expense of reduced agricultural trade... was the principal economic objective of Europe during the 1960s'.[1] The urge to believe that we do not know a workable general approach to our fundamental agricultural trade questions is overpowering. That exporting nations had excessive expectations in the 1960s with respect to Europe is now quite clear.

Important gains have been made in the international rules governing trade in industrial products, but there are still great problems there too, and some recent losses. In agriculture there continues to be widespread use of quantitative restrictions, export subsidies, special protective levies and non-tariff barriers of various kinds, especially rising subsidisation. Maintenance, and even increasing use, of these instruments is often justified on the grounds that special conditions prevailing in agricultural sectors – conditions relating to the difficult adjustment problems of peasant agriculture, the inability of industrial economies to absorb people from farm and rural areas rapidly enough and so on – justify certain trade restrictive actions. These practices are not limited to agriculture, however. There has been a pervasive tendency through twenty-five years of postwar experience to underrate the difficulties of human adjustment in the agricultural sectors of all developed countries, and to fail to provide for enough time to solve the problems of people. In the end, we have taken short-run and patchwork approaches to long-run agricultural production and adjustment problems nearly everywhere. These, in turn, have perpetuated situations that have determined agricultural trade policies and overall trade results.

The general opinion about domestic agricultural policies in most of the developed countries, especially Europe and the USA, is that they are unquestionably less tenable from a national interest standpoint than other economic policies governments have adopted. Some of this is nothing more than a search for scapegoats, or a failure to understand the crushing forces that are shaping the agricultural sectors of developed economies. When one considers the monetary mess that has gripped the world for many years, or the general failure of developed country attitudes and policies toward developing countries to achieve either rapport with poor nations or rapid progress for them, or the immense problems arising out of the multinational corporation and the movement of capital from one country to another, then the true nature and importance of the problems of world agricultural production and trade begin to come more clearly into focus.

We are not living in a world where all the economic problems except those in agriculture are being wrestled with seriously and solved. In most instances agricultural problems are not being handled with demonstrably less skill than other economic problems. When one puts the economic policies that major nations apply to their various sectors into perspective, agricultural problems do not loom larger than the others, only different. Their solutions may require longer time schedules and thus greater official patience than some other sectors. Agricultural protectionism, especially in Europe, is rooted in a slow-moving peasant agriculture, now changing as much in a decade as in whole centuries in the past, but still changing only slowly. Domestic agricultural adjustment and international agricultural trade are somewhat like a low-grade virus infection, always a nuisance and a slight drag on the general economic health, but seldom of crisis proportions that require heroic solutions.

The place of domestic agricultural policy

It is generally agreed that the question of expanding agricultural trade at a more rapid rate than in the past, if not at a maximum rate that would be possible under free-trade conditions in the next decade, is principally a question of restructuring domestic agricultural policies. The target is to reduce levels of protection in the developed countries that control most agricultural trade. The problem is how to do it. The conceptual and practical failures of the past provide relatively little guidance.

The typical academic recommendation has been to state that since domestic policies are the principal source of restrictions to trade, they must now become the subject of trade negotiations among nations. This is not entirely useless as a prescription for trade policy, but it has serious limitations. Negotiations among nations regarding agricultural policies probably offer more promise as a means of avoiding further trade restrictions than of speeding progress toward trade expansion.

There is probably no country in the world that is ready to submit its future domestic agricultural policies to meaningful international bargaining, with the objective of increasing the quantity and value of imports of agricultural products. The most careful advance political preparations are required at home for such an exercise, and no country has yet made those preparations. Adverse impacts of changes in domestic agricultural policies as a result of trade negotiations are greatly feared by powerful agricultural producer groups in all the developed countries. Thus domestic farm policies will not soon be decided, or even materially influenced, in the give and take of international trade bargaining.

If the objective of negotiating domestic agricultural policies internationally is too grandiose for practical consideration, what can be done?

First, the possibilities for accelerated trade expansion in agricultural products as a result of international negotiations should be assessed objectively, and without any binding advance commitment to lock ambitious objectives for agricultural products into the general round of trade negotiations scheduled to get under way in 1973 or 1974. The barely perceptible pace of change in the economic and political structure of world agriculture, and in the political possibilities for rationalising domestic policies in most countries, does not permit any other approach.

Second, objectives for marginal agricultural trade expansion should be established realistically. Both the volume and value of world agricultural trade are going to expand greatly in the 1970s, almost without regard to agricultural policies in Europe, Japan, the USA or Canada. Agricultural trade will expand sharply irrespective of the outcome of the prospective trade negotiation. The most protective inclinations and efforts of farm organisations and politicians are not likely to prevail against the powerful world-wide competition for access to the lands that are now being used for agricultural purposes, and the prospect of higher personal incomes. The near certainty that the demand for meat and other animal products will expand sharply is a key factor in the prospective expansion of agricultural trade. This will require importation of grains and oilseeds in record volumes by most of the developed countries for as far ahead as one can project.

Thus, in establishing objectives for increasing agricultural trade through our own efforts, we are dealing with a rather narrow margin of further expansion in an already expanding entity. It is far more difficult to get agitated politically over the necessity for even more rapid movement in already dynamic trade flows than it would be to develop international concern over stagnant or declining trade flows.

Self-interest as an alternative to negotiation

The political pressures and the rationale for trade-restrictive domestic agricultural policies have arisen principally out of the concept that certain agricultural pressure groups in various countries have of their own self-interest and that of agricultural producers. The rationale for political actions by governments, aimed at limiting the protective impact of agricultural policies and realising to a greater degree the benefits of further international division of agricultural labour by consumers, will have to arise out of a political assertion of self-interest by other groups. This is difficult and will probably take a long time. Urban and consumer groups are not highly organised in most countries, and making a direct challenge to agricultural policies is seldom at the top of their agenda. Governments often tend to be more responsive to organisations and traditions that have developed out of the past than to important political groups that are just emerging.

If a government is to follow the self-interest of the majority of its citizens in the matter of agricultural price guarantees, budgetary expenditures, food prices and trade, it must find ways of judging what that interest is. Little definitive study has been devoted to this question so far as agricultural policies are concerned. Yet there is a broad intuitive judgement among economists and the public that if all the costs and benefits could be weighed, the customary defensive responses by governments to agricultural price and income issues have cost more over recent decades and generations than remedial actions would have cost over a similar time-span. This kind of action would not represent a retreat from the long-standing public and treasury commitment to farm-people, but a means of compressing the adjustment and income payments into a limited time-period while helping farm-people and their children to shift into new lines of employment. Decisive agricultural actions of this type have seldom been taken by governments, except on a token or pilot basis, because the organisational and power base needed to prepare seriously for the future has seldom been as powerful as the groups organised to preserve the past.

Major trading nations will find themselves on rather different timetables over the next two decades in amending their domestic agricultural policies in a manner that would expand agricultural trade beyond the levels that will prevail under existing policies. This results from the differing stages of economic development and differing balances of political power in various countries. As a result, the time-honoured tradition of reciprocity in trade negotiations is not a very useful rationale for altering domestic agricultural policies of major countries, especially when it is applied largely within agricultural sectors rather than across industry as well. The underlying notion that imports are inherently bad for a country while exports are good, and that tariff reductions, or domestic farm policy changes that would have similar effects, by a country benefit other countries most of all is an insidious and self-defeating concept.

Transferred to the idea of altering domestic agricultural policies in ways that would expand trade, the desire on the part of countries for intra-sector reciprocity is simply a formula for not improving agricultural policies, even when over the short or long run it is clearly in the public interest to do so. In the future the USA, for example, should look at its price support level for milk and the associated quantitative restrictions on dairy-product imports, or at its subsidy payments to cotton and wool-growers, in terms of its own domestic interests as broadly conceived, and not in terms of what other countries are about to do in such matters. If it is in the public interest to rely more on butter imports than in the past because American farmers demand a very high price to produce enough to go around, we should not wait until Europe is in a political position to lower its milk-price guarantees or to announce a more favourable grain-meat price ratio, to make such a move in respect of USA milk prices.

Similarly, if the USA is to limit or terminate export subsidies for wheat, while preserving its competitive position in other ways, it should do so because, on balance, it is good policy and because it represents the way America

26

wants to spend public money in the years ahead. If other nations will take reciprocal action, so much the better.

By careful timing, it might also be possible to use the occasion of such unilateral actions on agricultural policy by one country to encourage other countries to take what would appear to be reciprocal actions, but would be, in fact, actions they were going to take in the interest of their own consumers when the proper occasion arose. Thus what would appear to be agricultural negotiations would actually be a series of agricultural orchestrations. Individual countries may want to package domestic agricultural changes made for domestic reasons in the tinsel of an international negotiation. The real impetus for constructive, trade-expanding moves on domestic agricultural policies in the 1970s and 1980s will usually be internal.

Various countries are in different positions with respect to the internal possibilities for improving agricultural policies in a way that would expand agricultural trade. Japan, despite its continued high level of protection on agricultural products, has for the past ten years given the world an important example of limiting its agricultural policies in the interest of import expansion without apparent reciprocity on the part of other countries. Japan embarked many years ago on a policy of increasing reliance on agricultural imports. This growing import demand arose partly from limited agricultural resources and partly from basic changes in the pattern of food consumption associated with higher levels of income. These income patterns seem likely to continue. Japan, with some difficulty, could probably have held the importation of cereals over the past ten years to lower levels and could have relied more heavily than she has upon food products produced at home. Instead she went the route of greater reliance on imports. According to a recent paper by Mr Michael Tracy,[2] Japan is now becoming reconciled to a situation in which agricultural pursuits by most cultivators will be a subsidiary activity, with income from non-agricultural sources being most important to the family. There is still much room for improvement in the structure of Japanese agriculture and in trade policy. Yet it is generally expected, because of increased demand for livestock products in Japan, that her imports of agricultural products will increase rather rapidly over the coming decades, almost without regard to what kind of agricultural policies Japan pursues, or what the rest of the developed countries do.

The EEC will one day find itself in a position similar to Japan's today. Demands for agricultural products will rise more rapidly than Europe's ability to produce them at reasonable prices, even with a rapidly changing agriculture. This will coincide with a rise in urban political power. The time for this is still some distance away. Europe has the technological capacity to increase agricultural production materially in the 1970s, so long as incentives for production are kept high. The theoretical potential market for agricultural products in Europe will tantalise, but probably disappoint, agricultural exporting nations in the 1970s and 1980s, as European governments and the EEC Commission find themselves unable to meet the political challenge required to turn increasingly to agricultural imports, while compensating for rapid agricultural change with the gains from food imports. The prospects for change would be enhanced if treasury ministers were having a strong influence on agricultural expenditures, but so far they are not.

The great problem that the rest of the world faces now with respect to Europe is to find the means, if any exist, of helping the EEC to avoid further increases in the level of agricultural protection as the Community expands in the 1970s. A great deal of momentum has been generated by eight years of increasing levels of protection under the CAP and by the decisions taken in connection with enlargement of the EEC. The political lessons of that period will not be lost on the more sophisticated and better financed farmers of the 1970s, including those in Britain about to share in the CAP for the first time.

The American role in furthering progress in agricultural trade in the decade ahead can take several forms. Like Europe, the USA is in some danger of falling back into old agricultural policy errors. Some farmers and their representatives still want 'cost of production', including a market return on the capitalised value of land and 'full parity prices'. Producers in the USA believe that world trading prices for grains are too low and want to find the means of raising them. Avoiding the error of a new and costly round of producer subsidies on export crops is one contribution the USA can make.

A second is to take some small initiatives toward increased agricultural imports. Dairy products and beef offer the best prospects and a start is being made on the latter. But the USA will move toward greater imports largely for its own reasons or not at all, and not because Europe will reform the CAP or lower the wheat price, or because Japan has reduced the duty on soya beans.

References

1) Schnittker, J. A., 'Reflections on Trade and Agriculture', in *Essays in Honor of Thorkil Kristensen* (OECD, Paris, 1970).
2) Tracy, Michael, *Japanese Agriculture at the Crossroads*, Agricultural Trade Paper No. 2 (Trade Policy Research Centre, London, 1972).

3 The Impact on Less Developed Countries of the Enlargement of the European Economic Community

Professor S. J. ROGERS,
Agricultural Adjustment Unit,
University of Newcastle-upon-Tyne.

The plight of poor countries and poor people, becoming relatively poorer at a time when the world on average is achieving unprecedented prosperity, is a major topic of international debate. There is a large and rapidly growing body of literature covering the many facets of a vast and complex subject. This paper is concerned with only the narrow aspect of the effect on less developed countries (LDC) of the EEC's domestic agricultural and international trade policies. Inevitably such a single-dimensional approach runs the risk of over-simplification, but it is hoped that it is sufficient to support the argument that the EEC's policies are of critical importance to some LDCs and should not, therefore, be decided on grounds of internal convenience alone.

It is worthy of note that the nations of the world have not only recognised the problem of the LDCs, but have agreed to co-operate to seek solutions. Thus:

> Governments designate the 1970's as the Second United Nations Development Decade and pledge themselves, individually and collectively, to pursue policies designed to create a more just and rational world economic and social order in which equality of opportunities should be as much a prerogative of nations as of individuals within a nation. They subscribe to the goals and objectives of the Decade and resolve to take the measures to translate them into reality.[1]

This general objective has, for the first time, been given some quantitative definition: 'The average annual rate of growth in the gross product of the developing countries as a whole during the Second United Nations Development Decade should be at least 6 per cent, with the possibility of attaining a higher rate in the second half of the Decade to be specified on the basis of a comprehensive mid-term review'.[2] There are some further consequential targets for savings, expansion of industrial and agricultural output; in particular, one of the critical factors is that to acquire the capital goods necessary to attain the growth targets and meet their debt obligations, the LDCs will require foreign exchange resources to grow by over 7 per cent per annum.

Foreign exchange is available to LDCs from three sources: export earnings provide about 75 per cent, foreign private capital about 10 per cent and foreign aid about 15 per cent. Hence, a fairly direct conclusion is that 'Trade is the main potential obstacle in the way of achieving the 6 per cent target of the second development decade'.[3] This importance of trade can be expressed more abstractly. Thus:

> The trade policies of the rich nations are directly relevant to the development of the poor nations in two different ways. The availability of trade opportunities can trigger off 'export-led' growth: trade then being the 'engine of growth'.... Alternatively this may make more rapid growth possible if export opportunities are opened up and thus the incidence of a 'structural' bottleneck (that is a 'foreign exchange gap') is reduced: trade is then 'permissive' and makes growth possible instead of initiating it.[4]

Of all the items exported by LDCs, mineral oil exports have been the most promising, expanding by about 10 per cent per annum until they now account for one-third of total LDC exports. However, as was pointed out by the President of the World Bank and its affiliates, three-quarters of these oil earnings go to six countries, containing only about one-fortieth of the population of the developing world (excluding mainland China). The other developing countries obtain about two-thirds of their export earnings from primary commodities and one-third from manufactures. Most of the primary exports are foodstuffs and agricultural raw materials. These are growing very slowly. For some items the market in the richer countries is not expanding very quickly, and for others, especially foodstuffs, the problem is aggravated by protectionist policies.[5] In addition to the search for expanded markets, the LDCs would also like stable markets, for widely fluctuating prices and volumes of trade persist in primary-product markets. While stability can become the modern version of the philosopher's stone, there can be little doubt that violent and unpredictable fluctuations present difficulties for the LDCs when trying to plan and finance development.

To recapitulate, if the proposed rate of economic development of the LDCs is not to be impeded by lack of foreign exchange, export earnings will have to expand by around 7 per cent per annum, preferably with more stability. For many LDCs this will necessitate an expansion of agricultural exports and manufactures derived from agricultural raw materials. Since exports by LDCs to the enlarged EEC amount to around 40 per cent of their total exports, and about half of exports to developed countries, the LDCs will have

30

to find an expanding market in the EEC to have a reasonable chance of achieving their goals.

To progress beyond these generalities, it is necessary to particularise with respect to country and commodity, for underlying the aggregate statistics are some strong bilateral trade links with a high degree of commodity specialisation in trade. Almost forty countries depend on the enlarged EEC for more than half their export earnings, while a further thirty obtain at least a quarter of their export earnings from the EEC. A cursory analysis shows that agricultural items have an important role. For example, of the countries with more than 50 per cent of their total export earnings from the EEC, about a dozen rely on agricultural primary products for more than 75 per cent of their export earnings. The comparable figure for the thirty countries with a 25 to 50 per cent dependence on the EEC is fifteen.

The market prospects in theory

During the 1960s LDC exports expanded at the rate of something over 6 per cent per annum. However, several qualifications must be made. To begin with, there is the point, already mentioned, that mineral oils accounted for a large part of the expansion. In addition, there is the general rate of inflation, and also the adverse movement in the terms of trade to be considered. Agricultural export prices may have increased by 10 per cent over the decade, but prices of manufactured goods increased by considerably more, so that the increase of 6 per cent in their exports did not permit the LDCs to increase their imports of the capital goods needed for development by a similar amount.

Instead of embarking on an elaborate forecasting procedure for the 1970s, a simplified analysis of the theoretical market prospects is made here. So far as agricultural products are concerned, three types of commodity can be distinguished. There are those that are entirely tropical, like tea or coffee, and thus non-competitive with EEC domestic production. There are those commodities, like sugar and oilseeds, that are at least partially competitive with domestic agricultural production in the EEC. There is a third group, important to particular LDCs, namely agricultural raw materials, which compete with synthetics such as man-made fibres or synthesised oils.

For non-competitive export products, if all other things stay the same, and this is discussed later, market expansion will come from rising population and rising income. If it is assumed that population growth will be around 1 per cent and *per capita* income growth about 4 per cent per annum, with an overall income elasticity of around $+0.25$ per cent, then the rate of market expansion is 2 per cent per annum in real terms.

The main difference in an analysis of market prospects for competitive products stems not from their demand characteristics, but from the fact that the protected domestic industry in the importing country often has the physical capacity to expand production more rapidly than the market, thus either depressing prices, replacing imports or both. The situation that arises under these circumstances is illustrated in Table 3.1. For ease of exposition, specific assumptions have been made. Demand is assumed to increase by 2 per cent per annum, price elasticity of demand is assumed to be −0.5 per cent, with domestic production accounting for two-thirds of total supply. The relative size of domestic supply will affect absolute changes in LDC exports. However, the key factor is whether the residual amount of market requirements less domestic production is expanding or not. Within this set of assumptions two changes are examined: domestic EEC production changes from an increase of 5 per cent per annum to a decrease of 1 per cent per annum; and LDC export quantities increase from zero to 9 per cent per annum.

The figures are indices. In the base year, the market price (P), the export quantity (Q) and the value of export earnings (V) are equal to 100. The figures in each cell show the indices at the end of year one, from which annual rates of growth can be derived. Thus in the first box, V equals 105.3 and the annual rate of expansion of LDC foreign exchange earnings is 5.3 per cent.

On these, admittedly simple, assumptions it becomes clear, from scanning the rows and columns of V, that the major variable affecting LDC export

Table 3.1

Effect of Market Changes on LDC Export Earnings

		LDC Export Quantity Change (Per cent per annum)			
		0	+3	+6	+9
EEC Production Change (per cent per annum)	−1	Q=100 P=105.3 V=105.3	Q=103 P=103.3 V=106.3	Q=106 P=101.3 V=107.3	Q=109 P= 99.3 V=108.3
	+2	Q=100 P=101.3 V=101.3	Q=103 P= 99.3 V=102.3	Q=106 P= 97.3 V=103.3	Q=109 P= 95.3 V=104.3
	+5	Q=100 P= 97.3 V= 97.3	Q=103 P= 95.3 V= 98.3	Q=106 P= 97.3 V= 99.3	Q=109 P= 91.3 V=100.3

earnings for these competitive commodities is not LDC export efforts, in terms of quantities offered, but the response of EEC domestic producers, to which can be added, for convenience without distinction, the response of exporters in other developed countries. Thus if EEC producers expand their marketings by 3 per cent per annum, which is about the rate experienced in the 1960s, LDC export earnings can, at best, only be maintained and would be more likely to be reduced.

There are bound to be some agricultural products with better than average prospects. Moreover, the theoretical prospects for non-agricultural raw materials and manufactures is somewhat better, in view of their more favourable demand characteristics. The situation for agricultural commodities that compete with synthetics is probably rather close to that of competitive products. Overall, in view of the dominance of agricultural trade for the non-oil-producing countries, it is difficult to resist the conclusion that the 'all things being equal' trade prospects are not good.[6] It is appropriate therefore, to discuss (a) how far the enlargement of the EEC affects the LDCs and (b) how far EEC policies could be altered to provide improved market opportunities for the LDCs.

Changes in LDC trading conditions following enlargement of the EEC

Enlargement of the EEC affects the trading conditions of each LDC differently, depending on its existing relationship with the EEC and the UK and the commodities of interest to it. Three main groups of countries can be distinguished.

Firstly, there are six countries that have a special relationship with the EEC. Simple trading agreements have been reached by the EEC with Spain, Yugoslavia, Israel, Egypt, Lebanon and Argentina; and others are in course of negotiation. There are formal treaties of association covering eighteen African and Malagasy states (Yaoundé Convention), the three East African Community states (Arusha Convention), Greece, Turkey, Malta, Morocco and Tunisia; again others are in course of negotiation. For this group of countries, enlargement of the EEC generally will bring an expansion of market possibilities. There is only one possible offsetting disadvantage that might apply to countries with formal association status. Countries with associated status normally have to grant reciprocal trading preferences to all EEC members; this practice is obviously regarded with disfavour by other developed countries. If enlargement of the EEC is sufficiently serious to provoke some retaliation (for example, it has been suggested that some developed countries' general schemes of preferences should not apply to LDCs with EEC associated status), there could be some offset of detriment to LDCs to put against the advantage of market expansion. At present, however, this does not seem likely and, in any case, such action could be averted if the EEC were to drop its insistence on reverse preferences, which are difficult to justify at the best of times.

Secondly, there are those countries that at present have a special relationship with the UK, either through Commonwealth preference or some commodity agreement. Some of these countries have been offered associated status, whilst others, for example the exporting countries (other than Australia) who are party to the Commonwealth Sugar Agreement, have received undertakings about market access. Some countries, however, have been categorised as 'non-associable', that is the EEC intends to exclude them from association, and this category includes, most importantly, Commonwealth Asian countries. In general, the EEC, as opposed to the individual member countries, has paid little attention to non-associated LDCs. Indeed the concept of association is that it should be restricted to a relatively small group of countries.[7] In consequence, Commonwealth countries can be split into those that will benefit from association, which may also provide benefits on the agricultural side, and those that will not. The latter, being mostly Commonwealth Asian countries, large in population and low in incomes, are likely to lose market access for items like jute, coir, handicrafts, cotton textiles and vegetable oils. Whether such a discriminatory policy can, or should, continue after the enlargement of the EEC is both complex and unlikely to be relevant to the 1970s during which present arrangements seem likely to continue.

The third group of countries comprises those that at present have no special relationships with either the UK or the EEC. The changes in market access for these countries following enlargement arises from the adoption by the UK of the EEC's Common External Tariff in place of its present system of tariffs. Of particular interest for LDCs, most developed countries have offered general schemes of preference; and those of the UK and the EEC differ from each other, not only in the items covered and the amounts of tariff remaining, but also in the basic approach. A brief examination of the two schemes suggests that the EEC scheme is complex, difficult to administer and on balance less liberal than the UK scheme, although particular countries may be differentially affected, particularly with regard to agricultural products and textiles and with regard to the possibility of trade expansion, which, of course, is where it counts. However, the two schemes have to be harmonised and whilst it is difficult to predict the outcome, it would require some optimism to foresee harmonisation leading to a major improvement in market access. It is prudent to assume, therefore, that, at best, this third group of countries will be in no worse a position with regard to market access than they are now.

In brief, the changes in tariff and market access consequent upon British

34

entry into the EEC are diverse. Some countries will benefit and others will lose, with the balance, if anything, being somewhat unfavourable for trade expansion.

Enlargement of the EEC is likely to lead to trade diversion, with agricultural imports to the UK likely to come from other EEC countries rather than other sources, including the LDCs. As a result, there may be some reduction in exports, most of which are subsidised, from the EEC to third countries. Providing total domestic production does not expand much purely as a result of the enlargement of the EEC, the net result of trade diversion on the LDCs may, if anything, be slightly beneficial, since their competitive position in third-country markets will be improved by the removal of subsidised EEC exports and this might more than offset the loss in European markets.

Some of the barriers to trade with the EEC are raised by means of national policies, an obvious example being an indirect tax or an excise duty. West Germany is reputed to raise five times more revenue from these types of duty on tropical products than it provides in its overseas aid programme.[8] There is, however, no reason for supposing that EEC enlargement will have any bearing on this aspect of market access.

Finally, as an effect of enlargement, the higher food prices in the UK may lead to a reduction in the total market for food, while the higher farm-gate prices in the UK, Ireland and Denmark could lead to expanded domestic supply so that the residual demand for imports will be smaller. While there is some substance in this argument, recent British governments have been moving away from the traditional policy of an open market for food, coupled with deficiency payments for domestic producers, to a system of import controls and levies. This means that, as far as temperate-food products are concerned, joining the EEC can be represented as an extension of British policy rather than a discontinuity. In so far as there will be a reduction in the import demand, this may relate to exotic products covered by the CAP, for example citrus fruit.

The purpose of this section was to review briefly the consequences of EEC enlargement, to discover whether one could expect there to be any improvement in the European market for LDC exports. The conclusion must be that, although a few relatively small LDCs may benefit, in general the prospects are not encouraging. It is not that the intentions of the EEC are ungenerous. Dr Mansholt, the President of the EEC Commission, addressing UNCTAD III in April 1972, stressed the need for a more liberal attitude towards trade with the developing countries of the world on the part of the advanced economies, especially the EEC. But inherent in some of the EEC's policies, both domestic and international, are factors that work against the intended goodwill. The following section, therefore, looks at some of these policies to explore what room there is for them to be amended in favour of the LDCs.

Possible EEC policy changes

The LDCs are looking for better market expansion prospects and this involves either obtaining higher prices from the same volume of trade, obtaining similar prices for a larger volume, or being able to put more value-added on to the products by processing. In their search for better expansion prospects the LDCs will be concerned with all products, but as has been pointed out earlier, many of them will be particularly concerned with agricultural products, both competitive and non-competitive with domestic EEC agriculture, and therefore with the CAP.

The new CAP structural policies are described elsewhere in this volume.[9] Whatever their general merits may be, they have little relevance to the LDCs, except in so far as they involve the complete removal of large areas of land from agricultural production. Restructuring agriculture into fewer, larger units can be expected to bring about little reduction in aggregate supply – on the contrary, the opposite may be the case. Nor is there any evidence to support the view that a restructured agriculture will tolerate lower levels of product prices. Despite the claims implied by Dr Mansholt at Santiago that the Community is attempting to reshape agriculture and to abandon protectionism, there can be little optimism in the LDCs about the new structural policies, unless considerably more emphasis is placed on land retirement.

The pricing policies of the CAP have two potential effects on markets for LDC exports, one operating through consumption and the other through domestic supply. Lower retail prices would encourage consumption; lower farmgate prices would discourage domestic production, provided the price reductions were sufficiently large. Few commentators doubt that CAP pricing policies are destined to change, but the enlargement of the EEC, by bringing to the UK the 'benefits' of the CAP, namely dearer food and the opportunity of contributing to FEOGA, has provided the EEC with several years' grace over the need for reform and the 1970s are likely to end with price policies not significantly different from those existing now. Even if some change in policy is countenanced, it would be difficult to predict that the revisions would improve the position of the LDCs.

An unpleasant consequence of the present CAP price policies is their tendency to destabilise world markets. Imports into and exports from the EEC are controlled to stabilise internal prices. At best this policy, by insulating a major market, leaves any instability in the world with a smaller total available market and thus greater proportional instability. At worst EEC trading actions can accentuate instability. While this policy affects both

36

developed and less developed economies, it is the latter that are less well able to accommodate fluctuations. An example has occurred in the sugar market: when world sugar prices were low there were net exports from the EEC; at the recent high prices an export levy has been introduced, thus discouraging trade. Dr Mansholt has advocated negotiations for multilateral commodity agreements, presumably with the objective of providing the LDC's with better and more stable markets, but it is difficult to see how such agreements, at least regarding competitive products, can be accommodated within the CAP pricing system without some rather unusual features with regard to storage and stock control. In another context, when discussing the EEC's general scheme of preferences, Professor R. N. Cooper has observed: 'Producers in developing countries may compete with each other to bid down the sales price...', and 'the real beneficiaries... will be the European importers lucky enough to get duty free quotas'.[10] The point is that neither statements of good intentions, nor choice of instruments, for example a multilateral commodity agreement, is sufficient to ensure that benefits flow where they are intended. Perhaps for non-competitive products the prospects for meaningful agreements are better. Certainly the pressure will be on the EEC to achieve something and some extension of market sharing seems to offer the best chances. However, even if successful, such schemes should not be expected to provide significantly higher prices for LDC producers. Neither the altruism of the importing countries nor the self-discipline of the exporters can be pushed too far and the best that can be aimed at is reasonable stability at modestly favourable prices.

Nothing hitherto in this section presages changes in price/quantity market relationships that might be of major benefit to the LDCs. There remain two further issues, that of discrimination by the EEC among the LDCs and that of adding value to primary products by processing.

As has been pointed out earlier, the EEC is discriminatory in its treatment of the LDCs. Moreover, those LDCs favoured under such conventions as Yaoundé and Arusha have to offer reciprocal advantages to EEC members. This, in turn, has prompted other developed countries to counter by offering concessions only to LDCs without special EEC arrangements. In consequence there is developing a 'trading bloc' philosophy and individual LDCs have to decide whether or not it is likely to be worth having special treatment from the EEC. Given the economic difficulties of multilateralism and the realities of international politics, it seems probable that some element of selectivity will persist, and the best that can be sought is a sensible and consistent basis for selection and some moderation of the degree of discrimination, with some favour shown to the poorest of the LDCs. In any case, this issue is mostly about the distribution of trade between the LDCs rather than about the global total.

The final topic concerns the possibility of adding value to agricultural produce in the country of origin, thereby increasing foreign earnings. Since the crux of the development process is industrialisation, the possibility of processing raw materials and exporting them is most attractive. In many cases, however, the market is restricted, either in a total sense of both raw material and processed product facing barriers to entry, or in the partial sense when the raw material may receive preferential entry terms over the processed product to the extent that it may be completely out of the question to process in the LDCs. For example, the figures in Table 3.2 relate to copra (the raw material) and coconut oil (the first-stage processed product). Some LDCs have preferential access to some markets, but for the rest, since the value added by oil-extraction is of the order of 10 per cent, the differential tariff rates virtually preclude any processing. The potential gain by the LDCs from the removal of such discriminatory tariffs is difficult to quantify, but likely to be substantial. Equally difficult to quantify would be the extent of damage to the processing industries in the developed countries, and they are the only losers, everyone else gaining. However, in so far as there is any room for manoeuvre in improving the general schemes of preference by widening the coverage, this is presumably high on the list of priorities.

Summary and conclusions

The basic message in this paper is simple. The development of the LDCs is a, if not *the*, major requirement of the 1970s. For such development to take place at an adequate rate, there needs to be capital investment and,

Table 3.2

General (non-preferential) tariffs in force in major
importing countries in 1969

	Copra	Coconut oil
USA	1.25 cents/lb	3 cents/lb
UK	10 per cent	15 per cent
Japan	0	10 yen/kg or 10 per cent
EEC	0	5–15 per cent depending on quality

Source: quoted in B. H. Davey, and S. J. Rogers, 'The World Coconut Market', Department of Agricultural Economics, University of Newcastle-upon-Tyne (May 1971).

therefore, an expansion of foreign exchange earnings, which in practice means an expansion of export earnings. For many LDCs this means expanding export earnings from agricultural products and from the enlarged EEC. The general prospects for such expansion are not good. Markets for agricultural products are slow-growing and competitive supplies are often capable of expanding at a faster rate than demand, so that the potential for imports may even be contracting. The act of enlarging the EEC from six members to ten has important implications for the LDCs since the conditions of market access will alter. Some of the detailed points have yet to be negotiated and in any case the fate of any particular LDC is a matter of individual calculation. Nevertheless it seems that, on balance, enlargement of the EEC of itself is unlikely to benefit the LDCs. In attempting to assess what policy changes might take place to the advantage of the LDCs, one is on treacherous political ground, but a cursory review suggests that there may not be much room for manoeuvre externally because of internal difficulties, notwithstanding the fact that the declared intention of the EEC is to help the LDCs during the 1970s.

Perhaps these somewhat pessimistic conclusions may prove unfounded and trade development during the 1970s may outstrip even the substantial development of the 1960s. Nevertheless, it might be prudent for the LDCs to re-examine the whole strategy of the second development decade under more modest assumptions to determine whether such a revision has materially different implications. Secondly, it is surely appropriate that the developed countries should re-examine their own goals, with particular reference to their protectionism towards agriculture and agricultural processing, which so often benefits but a few in the domestic economy – and these sometimes only in the short run – while it hurts many, both in the particular community, and almost all outside the EEC.

References

1) *An International Development Strategy for the Second United Nations Development Decade*, Cmnd. 4568 (HMSO, London, 1971).
2) *Op. cit.*
3) Lewis, W. Arthur, 'Objectives and Prognostications', in Ranis, G. (ed.), *The Gap Between Rich and Poor Nations* (IEA, Macmillan, Basingstoke, 1972).
4) Bhagwati, J. N., 'Trade Policies for Development', in Ranis, G., *op. cit.*
5) Robert S. McNamara speaking at the annual meetings of the boards of governors of IBRD, IFC, IDA, in Washington DC, in 1971.
6) McNamara, *op. cit.*, puts the prospects for market expansion of non-fuel primary products at 3 to 4 per cent per annum, compared with the overall UNCTAD target of 7 per cent.
7) See, for example, Tulloch, P., 'Developing Countries and the Enlargement of the EEC', *ODI Review*, 5 (1972).

8) Attributed to J. Danieau, EEC Commissioner, *The Times*, (19 May 1972).
9) See Chapter 4.
10) *The EEC System of Generalized Tariff Preferences*, E.G.C. Discussion Paper No. 132 (Yale University, November 1971).

4 Recent EEC Decisions on Price and Structural Policy

J. van LIERDE*
Commission, Brussels
of the European Communities

Introduction

The fun of predicting the future is increased by telling others about the results; any embarrassment arises only later if one is asked to explain the difference between the forecasts and actual results. Even this can usually be avoided by pointing out that it was the timing of the forecasts that was at fault and that it will take a little longer to work out than you had supposed. As one agricultural economist said: 'My predictions were not wrong; they have just not occurred yet.' The point is that accurate economic forecasting is difficult enough and although this paper is mostly about the next ten years, it is the directions of movement rather than the precise timing upon which discussion should focus.

The Common Agricultural Policy is seen by some people as the *pièce de résistance* of European integration policies. For no other sector of the economy has a common policy been developed so far as for agriculture. Indeed, in some ways. the CAP was a necessary precursor to any Common Market organisation, but despite its pioneering role and its central importance it is not without its problems. From the taxpayer's point of view, it is worrying that the CAP accounted for 80 per cent of the Community's budget for 1972 of around 4 billion units of account[1] (u.a.), to which must be added further expenditure on agriculture by member states. At the same time, consumers are not likely to be satisfied with a set of prices which – with the exception of Switzerland and Sweden – are the highest of any developed economy, particularly since such disadvantages on the consumption side are not offset by high producer incomes. Notwithstanding their increased productivity, farmers within the EEC receive an average income estimated to be between one-quarter and one-third below incomes of comparable professional groups.

One of the main reasons for the difficulties the CAP has experienced is the nature of the market and price policies that have been applied hitherto. Too often prices have been set at levels higher than could be justified economically. In addition, price relativities between competitive and complementary products have been set without due regard to these interrelationships. One under-

41

lying feature of the policies is that decisions have to be agreed by national politicians, who inevitably consider not only economic factors but a range of social and political implications.

Another important factor has been the lack within the CAP of a common structural policy, that is a policy under which there is intervention with respect to the volume of production and the number of people and other resources employed in EEC agriculture. Again political factors are responsible to some extent for this deficiency. When the CAP was being hammered out, farmers' organisations and their political representatives were hoping for policies that would produce direct and fairly immediate increases in farm incomes. Such hopes were thought more likely to be realised through a price policy than through a structural policy, since the latter is both slower acting and less clearcut in its effects. Once the CAP was operational, national civil servants came to realize that 'common policy' meant 'common decision-making'; this implied decision-making in Brussels with national agencies relegated to an executive rather than a policy-making role. Not surprisingly then, given that there was an agreed price policy in the CAP, there has been no great enthusiasm for a common structural policy and the proposals that were agreed in 1972 have undergone some three or four years of discussion.

Despite all the difficulties and impediments however, by the end of March 1972 the Council of Ministers had agreed on both price and structural policies, the negotiations about which culminated in a marathon session of more than one hundred hours. As a result it can now be stated that the CAP has entered a new phase. From now on, solutions will be sought in both price and structural fields. Prices have been set, much more than in the past, by taking account of economic relationships. Structural measures have been enunciated that should lead on the one hand to the modernisation of some farms and farm enterprises and on the other to an outflow of resources, particularly older farmers, under terms that recognise the economic needs but are socially equitable. It is hoped that price and structural policies will be complementary so that by 1980:

(a) the total costs of government support to agriculture will be substantially reduced, and

(b) the aspirations of both consumers and producers can be realised within
the framework of a policy synthesis which is reasonable to both parties

Price levels for 1972/3

In March 1972 the Council of Ministers decided upon the prices for agricultural products that would apply during the 1972/3 season, the prices for animal products dating from 1 April 1972 and for crop products from 1 August 1972.

42

The commission's proposals and their acceptability

In June 1971 the Commission proposed price increases for 1972/3 of between 2 and 3 per cent, although it should be noted that in arriving at these percentages the income shortfall occasioned by inflation had not been fully taken into account. Under pressure, including that from the European Parliament, the Commission revised its proposals and in the spring of 1972 suggested increases of 8 per cent, to be spread over the two years 1972/3 and 1973/4. This higher figure allowed for the likely increases in farm costs and the inevitable delay between the announcement of price changes and the resulting income improvement. With regard to the different products, the Commission argued that differentials should favour livestock products, especially meat, at the expense of arable crops, especially grain. Cereal production in the Six reached a record level of 76 million metric tons in 1971, while sugar is already in surplus although world prices have risen recently. In consequence, only small price increases were suggested on the arable side. There is however a deficit on beef, and imports in 1972 are expected to be about 600,000 tons, 14 per cent more than in 1971. The Commission proposed a 9 per cent increase in the beef price, with a similar increase for milk, due to the assumed beef–milk relationship. It was considered that a further increase in the beef price would depress beef consumption to an undesirable extent. To avoid this, while offering producers some further incentive, the Commission proposed to give subsidies to beef producers in respect of the investments that they make.

These proposals by the Commission were – as might have been predicted – greeted with neither enthusiasm nor unanimity. COPA, the European-level farmers' organisation, advocated increases of 11 to 12 per cent. The Economic and Social Committee, in which the various professional organisations are represented, preferred 9 to 10 per cent. The member governments also differed in outlook. Germany and the Netherlands wanted cereal price increases of at least 5 per cent to cover increased costs, whereas France considered 3 per cent more than enough. Belgium and Luxembourg sought an increase in milk prices of 12 per cent, while the Netherlands suggested something less than 4 per cent. Italy, while being generally in favour of increased expenditure on structural policy, was against any price increases, particularly for cereals for which she is an importing country; if there were to be price increases Italy wanted them to apply also to horticultural produce and olives.

The negotiations held by the Council of Ministers on the basis of the Commission's proposals, in the face of a range of opinions and interests, and after several seemingly abortive sessions, finally produced agreement. The decisions are summarised in Table 4.1. Prices were also decided for other commodities, including wine and tobacco.

As far as arable crops are concerned, the premium on rye distributed for bread will be maintained. The premium paid in Italy on imports of feed grains to compensate for higher wages in Italian ports will also be maintained.

The cattle and calf prices in the table relate to the period 1 April to 14 September 1972. Perhaps because of impending Italian elections, it was not possible to negotiate a more substantial immediate increase in price. However cattle and calf guide prices have been raised to 780 and 965 units of account per metric ton respectively from 15 September 1972 to 31 March 1973. In autumn 1972, the Council will also take a decision on the proposal for an extra premium for beef and the import system for calves and young store cattle. In addition, by this same date the Commission will submit a report on the further subsidies for beef, mentioned earlier. Dairy product prices overall will increase by about 8 percent, but only one-quarter of the total increase will be through butter, with three-quarters through skimmed milk powder. The subsidy for skimmed milk remains unchanged, but the subsidy on milk powder used in animal feeding has been increased, from 1 April 1972, from 130.00 u.a. per metric ton to 176.20 u.a. per metric ton.

One important aspect of price policy concerns the unit of account, its relationship to national currencies and the consequential complexities. It has been decided that if there is a change in the value of the unit of account, when expressed in terms of a national currency, then this will not give rise to any change in farm prices expressed in the national currency of the individual country. At the same time, as the exchange rate varies, so a system of compensatory payments will be introduced.

The last modifications in excange rates between the unit of account and national currencies took place in 1969. Three main points can be mentioned in connection with the 1969 crisis:

(a) On 10 August 1969, France devalued her currency by 11.11 per cent. Conversion of units of account to francs was undertaken using the new exchange rate and without further consequences;

(b) on 30 September 1969, the German government floated the mark;

(c) on 26 October 1969, the German government fixed a new exchange rate for the mark involving a revaluation of 8.5 per cent.

This last point is especially important so far as the practical consequences are concerned, for this was the origin of the system of compensatory levies. On 31 October 1969 the Commission allowed the German government to put compensatory levies on imports and to give export subsidies. These arrangements applied to products for which an intervention price existed as well as to products whose prices depended on those for which intervention prices are fixed. The levies and subsidies were fixed on a lump-sum basis.

Monetary policy brought difficulties for agriculture for a second time on 15 August 1971. This was the date on which the USA decided to stop the

EEC Price Decisions, 1972/3 Farm Year (Units of Account per Metric Ton)

Commodity	Dates applicable	Target Prices		per cent increase	Basic intervention prices		Per cent increase
		1971/2	1972/3		1971/2	1972/3	
Wheat	1/8–31/7	109.44	113.80	4	100.72	104.75	4
Barley	1/8–31/7	100.21	104.25	4	92.02	95.70	4
Rye	1/8–31/7	100.42	105.45	5	92.82	97.46	5
Maize	1/8–31/7	96.90	101.75	5	–	–	–
Hard wheat	1/8–31/7	127.50	132.60	4	147.90[a]	153.80[a]	4
Sugar	1/7–30/6	238.00	245.50	3	–	–	–
White sugar	1/7–30/6	–	–		226.10	233.40	3.2
Sugar beet (main price)	1/7–30/6	–	–		17.00	17.68	4
Sugar price (half fat)	1/7–30/6	–	–		10.00	10.40	4
Rape-seed (colza)	1/8–31/7	202.50	208.60	4	196.50	202.60	3.1
Flax (u.a./ha.)	1/8–31/7	110.00	135.00		–	–	–
Hemp (u.a./ha.)	1/8–31/7	80.00	115.00		–	–	–
Cotton seed (u.a./ha.)	1/8–31/7	70.00	80.00		–	–	–
Olive oil	1/8–31/7	1187.50	1247.00	5	–	–	–
Beef (live) (guide price)	1/4–31/3	720.00	750.00[b]	4	–	–	–
Calves (live) guide price)	1/4–31/3	942.50	942.50[b]	0	–	–	–
Milk	1/4–31/3	109.00	117.70	8	–	–	–
Butter	1/4–31/3	–	–		1780.00	1800.00[c]	1
Skimmed milk powder	1/4–31/3	–	–		470.00	540.00	15
Pigs (slaughtered) (basic price)	1/4–31/3	800.00	825.00	3.1	–	–	–

[a] Minimum price to producer.
[b] Applicable 1.4.72 to 14.9.72.
[c] Price of 1860.0 applies from 15.9.72 to 31.3.73.

45

convertibility of the dollar and Germany and the Netherlands decided to let their currency float. Two stages can be distinguished. First, since May 1971 compensatory levies had been introduced by Germany and the Benelux countries on the basis of foreseen revaluations with respect to the dollar of 7 per cent and 3.6 per cent respectively. The products to which these measures applied were cereals, pigmeat, milk and milk products, eggs and poultry, sugar, fish, wine, some products of the fats sector and some processed products made from agricultural products but not included in Annex II of the Treaty. The second stage started with the stabilisation of currency rates on 18 December 1971. For Germany, Italy and the Benelux countries a range was introduced under which currencies could fluctuate with regard to the dollar by 4.5 per cent. The amounts of levies and subsidies were revised when the farm prices were fixed in March and April 1972.

The parity difference in currency between Benelux and Germany will be accommodated as soon as Germany has increased her value-added tax by 1.8 per cent; that is when the revaluation of these currencies is formalised. From that time agricultural product prices in Germany and Benelux will be harmonised. However, the major problem is between Germany and Benelux on the one hand and France and Italy on the other. The existing differences are covered at present by border taxes and equalisation measures, but it is intended that these will be rationalised, probably over a three-year period, without materially affecting farmers' incomes. It is not yet clear how this objective can best be attained. One possibility is for France and Italy to increase their prices by about 1 per cent per annum; another is that the other four countries reduce their prices by a similar amount. At all events the Commission will have to submit proposals in this regard.

Effects of changes in price and price policy

The prices determined by the Council of Ministers are target prices, intervention prices and threshold prices. Market prices are determined by the interaction between demand and supply, which operates freely within certain limits. The lowest market price will be related to the intervention price, taking account of transport and other distribution and marketing costs, while the highest market price will be similarly related to the threshold prices. The actual price on the market will depend mainly upon the pressure of domestic supply and demand.

For some cereals, market prices are already close to the minimum, related to the intervention prices. Consequently the increases in intervention prices for 1972/3 will be felt directly by most farmers. A similar situation exists for sugar, where the 4 per cent increase in the intervention price is likely to be reflected in producer prices.

Market prices for dairy products during the autumn and winter of 1971 were substantially higher than the newly announced intervention prices, but during the spring of 1972 they were lower. Currently, production levels are expanding throughout the EEC, but consumption is, if anything, decreasing with higher butter intervention prices and lower margarine prices, so that the market is once again moving into surplus. Market prices are likely to be close to the minimum, but the increase in the intervention price will limit the reduction.

For beef the situation is somewhat different. The guide price has been increased by 4 per cent. Each week this price is compared with the reference price, which is an average market price in the EEC. When the reference price is less than or equal to 93 per cent of the guide price intervention buying takes place. When the reference price is lower than the guide price, imports have to pay not only the standard 16 per cent *ad valorem* duty, but also a variable levy. When the reference price is 106 per cent or more of the guide price the variable levy is abolished. At the time of writing (April 1972) the reference price was 112 per cent of the guide price, so no levy was payable. However, the reference price may well decrease in the coming months and since the guide price has been increased levies will be applicable earlier than would otherwise have been the case. Thus protection has increased and thereby, indirectly, farmers' prices will be protected at the lower end of the range when reference prices are lower than guide prices, but not at the top of the range. Since the Community has a substantial import requirement of beef, it is unlikely that market prices will move adversely in the near future, so that in consequence the recent price decision will not have much effect.

The overall effect of the EEC price changes on farmers' incomes will vary widely, depending on locality, farm size and pattern of production. Cereal and sugar-beet producers should receive higher product prices and hence higher revenues. Milk producers will get higher prices than they might have done, but possibly lower than last year. Beef producers may hardly gain at all and may lose if they feed cereals. Egg and poultry producers may lose because of the increase in feeding costs and the open-ended prices for the end-products. But even this level of generalisation needs qualification for the particular circumstance.

The effect of the changes in administered prices on actual consumer prices is also complex. Because of the higher intervention price for wheat, the prices of bread and other wheaten products are likely to increase slightly. On the other hand, consumers are already paying higher prices for beef and milk than is implied by the new levels, so that the immediate impact here will be negligible. For butter, the price is expected to move close to the intervention level in the near future. Consequently retail prices will remain somewhat higher than they would otherwise have been. The increase in sugar price, in practice, will have little effect on consumers. In fact, to restrict the extent to which higher world prices could be reflected in internal EEC prices an export levy has been proposed; this would be introduced whenever the export price is 2 units of account per 100 kilograms more than the internal price.

While most observers would agree that the 1972/3 price changes represent a move towards a more appropriate pattern of relative prices between the various livestock and crop products, it cannot be asserted yet that the absolute price levels are leading to optimum allocation of resources, in terms of either total market equilibrium or location of production. Production may still be stimulated in regions that are at an economic disadvantage, but it is possible that the new prices may at least have reduced the disequilibrium between livestock and crop products. Before further moves towards better resource use can be achieved through the price mechanism, it may be necessary for future prices to be little, if at all, higher than the lowest level asked for by any member government.

Common structural policy

The first steps towards a structural policy were taken as long ago as 1961, but at that time there was little pressure for action on this aspect of policy and no agreement was reached. By 1969 it had become clearer that price policies by themselves were unlikely to fulfil the objectives of the CAP, and the so-called 'Mansholt Plan' was published as a memorandum. It generated considerable discussion within the Community, including criticism, especially in Germany. In March 1970 the Council of Ministers accepted, in general terms, a set of proposals – known as the 'mini-Mansholt Plan' – that stemmed from the original memorandum. A year later the main objectives, principles and means for a common structural policy were decided by the Council and this was developed in detail for a meeting in June 1971, which led, after further amendment, to the decisions taken at the end of March 1972.

The main objective of structural policy was to be the improvement of the economic and social condition of farmers. In striving towards this goal it was necessary that

(a) the total costs to society of structural reform should not be too high; and

(b) any measures should fit into the general framework of production and marketing development and should contribute to the realisation and maintenance of market equilibrium.

There were four principles to be observed in any structural policy. Firstly, the policy would be Community-wide, but administered by member states. Secondly, it would be a voluntary policy and rely on farmers themselves taking the initiative. Thirdly, any policy must allow for regional differences.

Finally, there would be a financial contribution from the EEC budget towards the cost of implementing the policy.

Three different means of achieving structural reform were proposed. Firstly, measures would be introduced to increase the mobility of resources out of agriculture, especially labour and land. Secondly, there would be a reform of the structure of production. Thirdly, there would be an improvement in market structure. The Council of Ministers has accepted the first two means, but has not yet discussed the third.

Measures to encourage the outflow of people from agriculture

Member states have agreed to introduce legislation both to encourage people to quit agriculture and also to use any land thereby liberated either to help improve the structure of remaining farms or for non-agricultural purposes such as afforestation or recreation. This legislation will provide for an annual payment to those who are prepared to leave agriculture, who are between 55 and 65 years old and who engage in agriculture as their main economic activity. The maximum annual amount which will be financed by FEOGA is 900 u.a. for a married couple, both of whom work on the farm, and 600 u.a. for an unmarried person. It is also possible to obtain annual payments for other members of the family or hired workers if these are within the social security scheme. There is considerable discretion open to member states. The amounts of annual payments may be changed, or even compounded into a lump sum. Payments can be made for a younger age group, down to 50 years old in some cases. Within a country there may be differential rates for the regions. Finally, member states may decide to apply any part of the regulations, or not to apply them at all: thus they may make no payments whatever.

For the first five years of operation of this scheme there are special financial arrangements. FEOGA will finance:
(a) all eligible persons between 60 and 65 years of age;
(b) heads of household where the farm is no bigger than 15 hectares (37.5 acres);
(c) widowed and handicapped people where the disability is 50 per cent or greater;
(d) in those member states where the proportion of the economically active population engaged in agriculture is more than 15 per cent (that is in Italy and Ireland) eligible persons between 55 and 60 years of age.

There is also to be a subsidy per hectare on the land that the retiring farmer gives up, but this will be financed entirely by member states, not by FEOGA.

To be eligible for these two types of payment the person applying must be the head of the farm business and must be willing to cease farming. He must also surrender his land to one of three purposes, namely:
(a) selling, or leasing for a period of at least twelve years, to a potential

economically viable farm, or where this is not possible to any already viable unit;

(b) selling, or leasing for a period of at least twelve years, to a land bank;
(c) changing the use of the land to recreation, afforestation or some other use in the general interest.

Some precautions need to be taken to prevent the uneconomic spending of public funds under these headings. In particular member states are to avoid making annual payments to farmers whose holdings have been reduced in size in recent years. Thus farmers whose holdings have been reduced essentially during the last years would not be eligible for annual payments, except where the land had been expropriated or purchased in the general interest. Where land ownership is divided between different owners, member states must ensure that at least 85 per cent of the land should be surrendered to one of the three purposes given above.

Total expenditure by FEOGA on these measures designed to encourage the outflow of people from agriculture is expected to amount to 272 million u.a. over the first five years of the programme. This figure excludes the related spending by member states.

Measures to improve the structure of farming

The main feature of these measures is that they are selective and draw a distinction between three types of farmer, depending on relative income levels. The types are defined by reference to a parity income that is at least equal to the average annual gross wage earned by non-agricultural workers in the same region, taking account of the relevant comparison between farmers and other groups, or, for hired workers on a reference farm in the particular region, at least equal to the comparable non-agricultural wage. On this basis the first category, which can be designated 'viable', comprises those who have parity incomes or better and, providing they continue to develop, could be expected to maintain their position for the next six years. It is intended that these incomes be maintained through price policies that give price increases that are at least sufficient to offset any cost increases. Member governments are permitted to subsidise interest rates for any investments made by this category of farmer, provided that the farmer pays at least 5 per cent.

The second group, which can be thought of as 'marginal', comprises those farmers who at present receive below-parity incomes, but who might be expected to achieve parity over the next six years, together with those farmers who, while at present achieving parity, are at risk of falling below this level in the near future, the member states having to define what is meant by 'near future'. It is the responsibility of member states to define the limits of this category, for which four main aids are available.

Firstly, there is a possible interest-rate subsidy on investments made on the farm of up to 5 per cent for a maximum period of twenty years, provided (a) the farmer pays at least 3 per cent himself; and (b) the maximum investment on which subsidy is given is 40,000 u.a. per full-time worker where the farm is currently providing below-parity income and 32,000 u.a. per full-time worker where the farm currently provides parity income. For the time being no subsidy is permitted relating to investment in poultry and egg production, pending an examination being carried out by a working party due to report by the end of 1972. Investments relating to pig production will be subsidised only where the farmer can in theory produce 35 per cent of the feed required for the livestock. Purchases of beef and sheep qualify for interest-rate subsidy only when, after the development plan has been realised, receipts from cattle, sheep and milk account for at least 60 per cent of total revenue. Secondly, this group of farmers will be given priority in acquiring any land that becomes available from the retirement schemes discussed in the preceding section. However, interest-rate subsidies will not be permitted for land purchase. Thirdly, in cases where the farmer's existing assets and personal property are insufficient security for him to borrow the required sum for investments, guarantees can be given on his behalf. Fourthly, payment of 450 u.a. can be given for four years in return for record-keeping and accounts being kept in an agreed form.

To be eligible for support under these schemes a farmer must submit and work to a farm development plan that will run for six years, and possibly longer in some regions. It must be possible for the farm, after development, to provide one or two full-time workers (including the farmer) with parity income, working no more than 2,300 hours a year per man. At the end of the development plan as much as one-fifth of the income may be derived from non-agricultural enterprises, such as tourism.

The third category of farmer, which can be called 'non-viable', comprises those who are at present receiving below-parity incomes and who are likely to continue to do so. The Commission has made proposals that, subject to certain conditions, farmers between 45 and 55 years of age should be eligible for income supplementation, but these proposals have not yet been accepted by the Council of Ministers. Some measures have been agreed, however. For five years national subsidies can be given to this category of farmer, providing they are not at a more favourable rate than those given to the viable and potentially viable groups.

It is intended to prohibit, in principle, all types of aid other than those mentioned above, with two exceptions. First, it should be possible to assist the transfer of farm buildings and equipment onto farms that are potentially viable. Second special measures may be envisaged for regions where there is a danger of depopulation.

Considerable room for national flexibility exists in applying these measures. Member states may change the rates of subsidy or they may adopt only some part of the measures or, for some regions, they may apply none of the measures. The total cost to FEOGA of this type of support is expected to amount to 368 million u.a. over the first five years of the schemes.

Measures to improve the socio-economic information available to the agricultural population

The Council has approved a directive covering socio-economic information, the training of farmers and retraining for non-agricultural employment. The composition of socio-economic information centres is left to the member states, provided that it satisfies the basic tenets of the directive. For each post filled in the centres FEOGA will contribute a lump sum, designed to meet part of the costs of the organisation. FEOGA will also pay 25 per cent of the actual expenses of education and training and for the work of socio-economic counsellors, who will be helping farmers decide about their future careers. It is expected that the first five years of these programmes will cost FEOGA about 92 million u.a.

Measures to improve market structure

The Council has not yet taken any decision on the Commission's proposals on producer groups and their unions. The main problems still to be resolved include: the field of application (for which products groups can be formed); who can be members of the groups (only farmers or also processors and/or traders); whether or not member states should be obliged to pay aid to recognised groups and unions; the functions of the unions (whether they should be simply informative or have the possibility of marketing produce). It has been agreed that the Council should decide by 1 October 1972 about the matter of producer groups and also on a range of topics covering long-term contracts, vertical integration and the reorganisation of firms in the processing industry. There is also the possibility of providing better market intelligence through the creation of a European marketing committee drawn from different professional groups. Proposals on these latter subjects will be laid before the Council in autumn 1972.

Effects of the new structural policies

Since the policies that have been agreed are novel and since member states have considerable freedom to implement the policy in different ways, any discussion of the effect of these policies must be extremely tentative. However, some qualitative observations can be made. Many farmers, especially the young, who may have financial difficulties or who may be on the margin of viability, will have the opportunity to obtain investment funds through subsidised credit and to acquire additional land. If these policies continue to operate effectively for some time it can be expected that the future structural pattern of EEC farming will be more evenly distributed, with few very small holdings remaining. These policies do not, however, facilitate the entry into farming of the farmer's son.

The retirement assistance should accelerate the rate of out-migration, particularly among the over 60-year-olds, although the extent of this will depend on the actions of member states. This will affect production, probably increasing it, and, to the extent that the people are not replaced in farming, will also lead to increases in average incomes by reducing the number of heads among which sectoral income has to be divided. Finally, since there will be some degree of harmonisation of policies as member states adopt instruments that are consistent with the new decisions and abandon existing different policies, there should be a reduction in the imperfection of competition.

Turning to the statistical implications of these policies, there are at present 4.2 million farmers, that is people who have at least half their economic activity in agriculture, in the EEC, with an age-distribution as follows:

Age group	Number of farmers (,000)
Below 45 years	1,250
45–54 years	950
55–64 years	1,100
65 years and over	900
Total	4,200

Of this total, about 200,000 are well into the viable category. With a further 500,000 more or less at that level, there are some 700,000 farms in the first category. This leaves some three and a half million farms in the marginal and non-viable categories.

It is expected that between 1972 and 1976 about 300,000 farmers will try to benefit from the annual payment schemes, of which 200,000 will be in Italy, where people in the age-group between 55 and 65 years are eligible for assistance, as against the 60 to 65 year age-group in the other member states. In addition, about 500,000 farms will be modernised, the total cost of this modernisation alone (excluding aids for book-keeping, groups, amalgamations) being estimated at 820 million u.a., of which FEOGA will contribute some 205 million u.a. Furthermore, about 2,000 socio-economic counsellors will be engaged and about 120,000 farmers will receive further training in farming.

It is difficult at this stage to make further estimates as to how many farmers will quit farming, benefiting from the structural improvement premium to be paid by member states. By the end of the first five-year period it is expected that about one-seventh of the farms that today are economically non-viable will benefit from the measures put forward for the development of farming. About one-tenth will benefit from measures financed by FEOGA and directed towards people who want to stop farming. If one adds to this an estimate of another 8 or 10 per cent that will quit farming on the basis of national measures, one comes to the conclusion that about one-third of non-viable farms will benefit from the various structural policies between 1972 and 1976. In addition, a large number of farmers may benefit from the measures to provide socio-economic information. This means, however, that by 1976 two-thirds of the farmers in the non-viable category will not have benefited from the new policies. It was with this type of reasoning in mind that the Council decided to review the policies and their effectiveness at the end of the first five years, with a view to making recommendations for the second five-year period, it having been agreed initially that some form of structural policies should run for ten years.

With regard to the patterns of production on the farms still in existence at the end of the ten-year period, it can be forecast that most of these will have become specialist units, although there is still likely to be a reasonable number of family or large mixed farms and some part-time holdings. The basic pressure favouring specialisation is technology, where the modern systems are highly demanding of both capital and management and can be practised only where there is a sufficient volume of production to spread the overheads. This pressure will be reinforced by most of the aids that have been discussed above. For example, the beef and sheep investment assistance is envisaged only for ruminant livestock specialists. However, in the case of pigs the new proposals will favour the farm that provides its own feeding stuffs, and these might be mixed. Complementing these changes on the production side will be changes in marketing, where more contract farming and vertical integration can be expected.

Notwithstanding the pressures towards specialisation a large amount of mixed farming will remain. Some farming systems, which are both well tried and reasonably rewarding, are based on complementarity between enterprises, for example for rotational reasons, while in some regions specialisation is virtually impracticable. Furthermore, there will still be some farmers who prefer to spread their risks and keep a range of enterprises going. Part-time farming, not too far from urban employment opportunities, will continue producing commodities that are not labour or management-demanding

54

Finally, there is a growing emphasis on the costs and disadvantages of environmental pollution and it can be expected that this will translate itself into policies that favour the dispersion of at least some types of production away from major population centres.

Conclusion

The Community's agricultural policies with regard to pricing and structure, which were formulated in March 1972, mark something of a departure from lines that had been followed earlier. An attempt has been made to put prices and price-relativities on a more rational footing with respect to the economic forces that are in evidence. However, recognising that price policies alone cannot achieve the objectives of the CAP, a new set of structural policies has been devised to help achieve the goal of transforming the structure of the industry over the next decade.

As yet these policies are incomplete. Nothing has yet been agreed on reforming market structures, nor have measures been proposed to prevent some later subdivision of a farm that at present is economic. Even where the policies have been formulated member nations still have the task of implementation and have retained substantial individual freedom of action, so that different patterns of change may emerge as a result of the variation of national policies. In any case structural policies are not easy to agree or to administer.

The policies that have been proposed are not costless. However the new approach offers the long-term advantage that any spending on structural policies should enable offsetting reductions in future support to be considered so that at the end of ten years the Community's agricultural policy should display more satisfactory cost characteristics than would otherwise be the case. It is hoped, and indeed it is the intention behind the new policies, that these new ideas may in the long run provide a situation that is satisfactory to all the parties concerned–producers, consumers, taxpayers and third countries – and it is by this criterion that the policies should be judged.

Reference

1) The unit of account was the value of the pre-devaluation US dollar. At present, approximately £1 sterling equals 2.40 units of account.

Note

* The views presented here are solely those of the author and do not reflect the views of the Commission of the EEC.

5 Problems of Adapting UK Institutions in the EEC

Professor D. K. BRITTON
School of Rural Economics and Related Studies,
Wye College.

Introduction

The studies that have been made in many quarters to try to determine the likely impact of Britain's entry into the EEC on her agriculture have been mainly concerned with questions of profitability, costs and security of operation for producers and traders. This paper is concerned with the institutions that serve British agriculture and its related industries, and the way in which they are likely to have to adapt themselves to meet the new situation.

It is natural that at the present time, with the date of formal accession drawing near, there is a busy preoccupation with the immediate problems that have to be settled if the UK is to conform to all the regulations that will begin to apply in this country, either in temporary or permanent form, from the date of accession. It is true, of course, that where prices are concerned we have the breathing space provided by the agreed transitional period, so that in some respects adaptation does not need to be instantaneous. Nevertheless, the sheer bulk of administrative detail is such that it would be understandable if for the time being it filled the horizon and obscured from view some of the long-term considerations.

We tend to think that the UK's task on entry is to modify her arrangements in conformity with a common agricultural policy that has been worked out without her participation, and in this sense there may be a feeling that her institutions have no positive role in the formation of policy at the present time. It is, moreover, deeply ingrained in the British national character that new entrants to a 'club' do not immediately announce their intention to campaign for a change in the rules. However, it is now widely recognised among the institutions of the EEC, as well as outside them, that the CAP is very far from being in a permanent form. It has evolved and it will evolve. Enlargement of the membership must in itself bring about some changes. The future evolution of the CAP will be a process in which the UK will have a significant part to play, and it should be emphasised that 'evolve' is a transitive as well as an intransitive verb. She is not simply at the receiving end of a set of regulations or directives that have set once and for all the pattern of European agricultural development.

In reviewing the various British agricultural institutions one must look at the ways in which they can prepare themselves to take part in the evolution of policy as well as the more immediate changes that they have to make in their day-to-day organisation. Nor can one ignore the point that difficulties may arise in reconciling long-term objectives with short-term expediencies. Certain lines of action seem to be necessary for virtually all the UK agricultural institutions, both governmental and non-governmental. Therefore, rather than dealing with the changing role of the institutions one by one – Parliament, the government departments, the marketing boards, the farmers' unions and so on – reference is made to one institution or another by way of illustration of the various kinds of adaptation that are likely to occur in them all, to a greater or lesser degree. In this way the omission of any particular organisation from specific mention should not be interpreted as a failure to recognise its due importance in the anatomy of British agriculture.

There are five ways in which institutions will have to adapt themselves to the new situation. They have to consider how they can organise themselves most effectively to inform; conform; perform; reform; and transform.

The need to inform

The first necessity that all institutions have to meet is to inform and advise their own membership about the facts of the CAP and their implications. This is basically a matter of assisting their members to make the transition to a new system and a new market situation as smoothly as possible and to advise them on the best way of meeting the requirements of the new regulations. People need to know whether they will be able to continue much as before or whether they face entirely new problems and risks. Many organisations have already gone a long way in this direction. The NFU has published its valuable *Farmers and Growers Guide to the EEC*; the Home-Grown Cereals Authority (HGCA) has issued a number of background documents dealing with marketing arrangements for cereals in the Common Market; the Economic Development Committee for Agriculture is preparing a series of studies dealing with the foreseeable competitive situation for various groups of commodities. Numerous other organisations have performed a similar task in their respective spheres. Sometimes the advice and assistance offered is very specific and technical. For example, the British Association of Grain, Seed, Feed and Agricultural Merchants (BASAM) has requested the Institute of Corn and Agricultural Merchants (ICAM) to arrange courses of instruction so that merchants can prepare themselves for the rigorous requirements relating to the quality standards that have to be met if grain is to be acceptable for intervention buying.

58

Traders' organisations can help their members to cope with the vast number of regulations that will affect them and to keep abreast of the changes that may occur from day to day. Without such help the paper-work of 'familiarisation' could well become a crippling burden, especially for the small firm.

Besides giving information and advice of this kind, organisations need to encourage their members to express their views about the various regulations, so that if representations have to be made for their modification or abolition, or for the formulation of new regulations to meet special circumstances, this can be done effectively. Thus a two-way flow of information needs to be established within organisations.

There are important new responsibilities in the field of statistics. Because of the new trading situation, statistics of European prices, of costs and of quantities produced and imported or exported will take on new significance for British farmers, growers and traders and they will need to be reliably reported and wisely interpreted. A sensible division of effort and responsibility between governmental and non-governmental statistical services should be achieved, with full liaison with the EEC's Statistical Office. International statistical comparisons must be improved and extended. The recent NEDO report on cereals contains an interesting example that may be typical of the kind of new development we may expect, where it presents a table indicating not only the producer's price of wheat in different EEC countries but also the estimated costs of production under 'average' and 'good' conditions and consequently the margins received by the farmers. No doubt there is much room for improvement in such estimates, but a beginning has been made.

In some ways Britain's membership of the Community will in itself impose a need for new kinds of statistical information to be sent to Brussels. The Meat and Livestock Commission (MLC) and the HGCA, for example, will have to provide the information on prices that is a requirement of all member countries. The government agricultural departments will have to supply farm accounting data from a representative sample of farms, in a form resembling that of the present Farm Management Survey in England and Wales (and corresponding enquiries in Scotland and Northern Ireland) – which provides material for the Annual Price Review – but also including new items, such as an estimate of the annual depreciation of buildings and fixed equipment, which have not hitherto been available in this country.

Another kind of information that will now become important is concerned with the interpretation of EEC regulations and of the provisions of the Treaty of Rome. The UK has to find out how far she can go in the direction of market management, vertical integration, centralised organisation of producers, or other ways of seeking market strength through collective action without 'distortion of competition' in a way that would be deemed incompatible with Common Market principles. The functions performed by the milk marketing boards are a case in point. At the time when the House of Commons Select Committee on Agriculture was taking evidence on the effect of entry on British agriculture, Ministry of Agriculture officials commented that the Commission favoured voluntary co-operatives rather than compulsory and statutory organisations for marketing, and they could not be sure what the position of the boards would be. After the negotiations of 1970/1 the Prime Minister reported to Parliament that the milk marketing boards, like the other marketing boards, were 'expected to continue their essential marketing functions'. More recently, the Minister of Agriculture has confirmed that 'the Board would be able to ensure that it got the best possible return for its members. It would be able to pool its financial returns and remunerate its members as it wished.' This kind of assurance and interpretation is likely to be called for in many other areas that are the subject of Community regulations. One development waits on another, and delay in announcing decisions can result in a long chain of frustration. Thus, until the methods of market intervention that will operate are fully specified, explained and understood, organisations, firms and farmers cannot plan their actions or assess their positions. There will also be a need to spell out clearly the respective areas of responsibility of the Commission, the national government and its appointed agencies in administering the various instruments of the CAP.

The clarification of regulations is not only a matter for Britain and the other new entrants to EEC. In the Six also there is uncertainty about the legality of some practices, and a kind of 'case law' is in process of development. At the present time the Commission has under examination hundreds of ways in which aid is given to agriculture by the governments of the Six member countries, and it intends to pronounce upon their compatibility with the regulations. Up to now, however, the criteria have not been clearly defined. In any case, the Commission has not found it easy to enforce its decisions, and the ultimate sanction of proceedings before the Court of Justice has very seldom been invoked. No doubt a fair degree of harmonisation will be achieved in due course, but meanwhile there appear to be large areas of uncertainty, which governments and other organisations may have a duty to dispel.

Personally, one can only deplore the attitude that found expression not long ago in a television programme in Britain when the speaker said, in all seriousness and without challenge from his interviewer, that it was a matter of urgency that farmers and civil servants should be devoting all their energies to 'finding ways of fiddling the rules the same as the Germans and French do'. British farmers and traders will generally be able to compete effectively in the sense usually meant by 'fair competition'. Do they really want to establish a reputation for outsmarting their European neighbours in bending the rules? However, this is not to say that there is an obligation on them or on anyone else to adhere strictly to regulations that are recognised by all concerned to be unworkable and to have fallen into abeyance. Britain should call for the abolition of such regulations. Meanwhile there is much work to be done in clearing up the 'grey' areas where harmless non-compliance merges into fraudulent contravention. Her agricultural institutions will need to keep their legal advisers active and vigilant.

There are other matters on which organisations need to keep themselves informed at their headquarters, without necessarily having any obligation to pass on all the information to their members. For instance, it is obviously important that government officials should be fully conversant with the workings of FEOGA and should not overlook any ways in which the UK might benefit from this Fund by qualifying for various forms of aid that might be applicable to it, such as aid to specially difficult regions. Similarly, they must know intimately the functions of the numerous European committees and the personalities involved in them. Representatives of farmers' and traders' organisations will need to keep in constant touch with the officials who serve on the management committees that advise the Commission on the various commodities and the regulations pertaining to them. This kind of information is indispensable, bearing in mind that organisations are engaged in influencing opinion in two areas simultaneously, one at the Community level and the other at the national level.

Before leaving the question of the need to inform and to be informed, some reference should be made to the vital role of television, radio and the press. If the orientation of British agricultural thinking has already shifted significantly towards Europe it is largely as a result of the way in which these media in Britain have responded to the need. They have developed their own links with their European counterparts and we should expect an even better coverage and depth of analysis in future, with a corresponding muting of emphasis on the purely British point of view.

The need to conform

During the transitional period Britain's institutions have to adapt themselves so as to conform to Community regulations in whatever ways this is required of them. It is a transitional period for institutions as well as for prices. Many of them will have to learn how to perform new duties. For instance, the MLC will eventually cease to be engaged in fatstock certification, but it will have to accustom producers and traders to the EEC schemes of carcase classification and grade descriptions. It will also take on new responsibilities as a price-reporting agency to the Commission. The HGCA will have to operate regional arrangements for the operation and surveillance of denaturing of grain, support-buying under the market intervention regulations, storage and selling out of store. It too will have to supply Brussels with price statistics and market intelligence. The new Intervention Board for Agricultural Produce (IBAP), responsible to the ministers of agriculture but not a part of the ministry, will have overall executive responsibility for the application of the market intervention regulations to Britain. Its expenditure will be subject to the oversight of the Comptroller and Auditor General and its administration to the scrutiny of the Ombudsman. The Board will be able to consider individual cases where hardship or injustice might arise. Its chief executive will be a civil servant, but other members of the Board may be drawn from commercial and other fields.

As was indicated in the memorandum accompanying the European Communities Bill, Parliament has made the legislative changes necessary:

(a) to give the force of law in the UK to present and future Community law;

(b) to provide for subordinate legislation so as to implement Community obligations or exercise rights under the relevant treaties;

(c) to provide for the UK's share of the Community budget and other dues;

(d) to provide for the charging of customs duties in accordance with the treaties;

(e) to set up the IBAP already mentioned;

(f) to amend certain laws, for example relating to seeds, food, feeding stuffs, animal health and plant health, so as to comply with obligations that arise from accession or will arise thereafter.

This is the framework for conformity to the CAP. Many points of detail have still to be negotiated. All this preparation has required considerable redeployment of staff in the Ministry of Agriculture, Fisheries and Food, to strengthen, in their EEC work, both the commodity and external relations divisions. Further staffing problems arise from the necessity for Britain, on entry, to provide its quota of the staff of the Commission of the enlarged Community, located chiefly in Brussels. This is not a matter of secondment in the usual sense. Commission staff may recognise no obligation or commitment to a national government and must therefore cease to be national civil servants for their period of service to the Commission. It seems inevitable, therefore, that Whitehall will have to surrender expert staff whom it can ill afford to lose, especially in these early days of continuing negotiation and alignment. In practical terms this may prove to be one of the major 'institutional adaptations' that will be called for in the next few years.

Conformity rather than formulation of new policies is likely to be the major preoccupation in the immediate future. This will no doubt be true of non-governmental institutions as well. For instance, it would hardly seem realistic for the NFU membership to press its officers to propose new marketing boards for cereals and meat, when the compatibility of the existing boards with Common Market regulations is still in some doubt. A question frequently asked is whether conformity with the CAP will lead to the abandonment or modification of the national Annual Price Review procedure. Although a number of commodities may be removed from the schedule appended to the existing Agriculture Acts for purposes of price guarantees – since these will no longer be operative at the national level – the Annual Review seems certain to continue. One of its main purposes has always been to enable the government, after consultation with the farmers, to strike a proper balance between the various interests concerned – the farmers, the consumers, the taxpayers, Britain's trading partners, and so on. Ministers will still need to carry out this exercise in order to decide upon the policy to be pursued in the Council of Ministers in Brussels when they confront the proposals on prices and other matters put to them by the Commission. They will have to arrive at their own assessment of the case which the British farmers' leaders will no doubt put to them for changes in European prices, and the Review will help them to determine where the shoe really pinches and what kind of price adjustments could reasonably be urged. The difference is that decision-making will now be one remove further from the reach of the national agricultural organisations. Conformity to the CAP, then, need not mean the disappearance of the Annual Review as a British agricultural institution, but it seems likely to become more of a briefing session and less of a negotiation.

The need to perform more effectively

Entry into the Community and acceptance of the CAP certainly present a challenge to our agricultural institutions to fulfil their tasks as efficiently as possible; that is, to attune themselves readily to the new circumstances and to make the most of the mechanisms that have been set up. Having informed ourselves of the situation, we have to act on the information as intelligently as we can. For instance, we have to take full advantage of EEC provisions to assist from Community funds the development of producers' groups, changes in agricultural structure by farm amalgamation encouraged by early retirement of farmers, measures to improve special areas, and so on.

There is much evidence that British agricultural organisations have now fully realised the need to affiliate with their European counterparts for more effective action. The closer links with Europe, both formal and informal, require Britain's institutions to explore and improve their channels of communication with Europe, and particularly with Brussels. This is necessary not only to ensure that they are supplied promptly with the information about developments, proposals or prospects that are likely to affect them intimately, but also because of their urgent need to discover which channels of communication towards Brussels can be used most effectively at the present time, or would be potentially of effective use in the future, so that their opinions can be voiced and their pressures applied in the most telling fashion. A number of developments are already to be observed under this heading. Many organisations, from Whitehall downwards, have taken special steps to ensure that selected members of staff are given language training to enable them to communicate personally and directly with their colleagues or opposite numbers in the other member countries. Naturally pride of place is being given to the French language, but German and other languages are not being ignored. Those who are closest to the centres of power in Brussels are those who emphasise most the necessity for this language training.

Secondly, under the same heading, we may note that a number of organisations now have their permanent representation in Brussels. The British government has its permanent delegation with one person of ambassadorial rank and two at the ministerial level. More recently the NFU has opened its Brussels office, with a director in residence there. The CBI also has an office in Brussels, and there is no doubt that a number of other institutions are pondering the balance of costs and benefits of following these examples. It is quite possible that if they do so a wasteful duplication of facilities could arise, and among the many tasks of co-ordination which confront them this is one that will need to be sorted out in a spirit of give and take.

Thirdly, numerous opportunities, and indeed obligations, will arise for institutions to send representatives from time to time – and often for considerable periods – to represent them on various committees. Some people have already been serving in this way for a considerable period, and particularly since the Six extended to the four applicant members the right to be consulted about all major new developments during the months immediately prior to entry. These committees may be at the official level – that is involving representation of the British government – or they may belong to non-governmental organisations such as COPA. All these initiatives involve the reallocation of scarce resources of highly qualified staff and there are obvious limits to the number of people who can be spared for such duties. Especially when progress in these committees is slow and they tend to get bogged down in a mass of details, the question is bound to arise whether these members of staff, who have other work calling for their attention in the UK, might not be better employed back home. It may take some time for the opportunity cost of systematic and continuous representation in Brussels to be assessed, but undoubtedly the traffic in the European direction is on the increase and seems likely to multiply further. This increase in European contacts has already caused many British agricultural institutions to rethink their roles and reshape their activities. The EDC for Agriculture has its Common Market Sub-Committee and has reorientated its whole work programme, recognising a continuing responsibility to assess the changing competitive situation, to seek new opportunities for British agricultural expansion and to make international comparisons of productivity and performance.

Apart from external reorientation and affiliation, there is a need for an improved internal relationship between our multifarious agricultural bodies. In the first place, if they are to avoid duplication of work and facilities in their endeavours to achieve proper representation in Brussels, they need in some measure to pool their resources and co-ordinate their actions. In this context the recent formation of the European Liaison Group for Agriculture (ELGA), which has brought together representatives of twenty-one UK farming and trade organisations under the leadership of the President of the NFU, is a step in the right direction. In the second place, it seems essential for the many relatively small producers' marketing groups and co-operatives to be drawn together under some kind of 'apex' organisation that could act as a main channel of communication with Europe on a multi-commodity basis and could harmonise the activities of the marketing boards. Perhaps the embryonic British Agricultural Marketing Development Organisation (BAMDO) will prove to be such an organisation, supported by regional or second-stage groupings of co-operatives and other groups that might be able to regulate effectively the volume and quality of production without prejudice to the interests of farmers in other member-countries of the Community. As the NFU has stated, 'the need for effective intervention now transcends all other demands which the farmer can make on the organisation to which he subscribes for support and assistance'. But, parallel with the debates on national sovereignty that have taken place in Parliament, there is a debate still to be resolved between those who value above all else their independence of action and those who are willing to sacrifice part of it in the interests of more effective co-operation. British entry is going to hasten a process of streamlining and mutual adaptation that would still have been necessary had she not joined, but would have taken much longer to bring about.

Improved performance by agricultural institutions through co-ordinated effort by groups that have hitherto worked independently, and even in rivalry with one another, will require international as well as national initiatives. Mr Tom Cowen, the NFU's Director in Brussels, speaking of the need for international co-operation in the agricultural sector, has said: 'I cannot believe that there is any great difference between the British farmer and his colleagues anywhere in Western Europe. Our aims, our hopes, our ideals, our difficulties, must be the same…. It behoves us for our future good to be a little less wrapped up in the mystique of our own uniqueness.'

The need to reform the CAP

Among farm organisations in the Six there is dissatisfaction with some features of the CAP, and within the Commission itself doubts have been expressed as to how long some of these features can last. British agricultural institutions must therefore be ready not only to adapt themselves to the CAP but to adapt it to the realities of the changing situation, remembering that European agriculture is itself going through the most rapid transformation it has ever seen.

The enlargement of the Community from six to ten members must itself give rise to changes in emphasis and a re-examination of priorities. This seems likely to proceed pragmatically, by driving a nail where it will go, rather than by any root-and-branch removal of existing arrangements, and a beginning will probably be made through the commodity approach that is so apparent today at the national and international level. The management committees in Brussels, composed of commodity experts, will be under pressure to bring about adjustments in prices and in other marketing arrangements where these have been shown to be inappropriate. For example, they may be involved in discussions about changing the price differentials between different kinds of grain or the regional differentials in intervention prices, if these do not correspond to economic realities. Within their own commodity areas they may find it worth while to concede points for the sake of harmonisation because this will lead in turn to a liberalisation of trade which at present is unduly restricted by the regulations.

Not all reforms, however, should find their expression through the commodity approach, which by its nature seems to be limited in its possibilities and even dangerous in its narrowness of vision. Certain strategies are precluded if the commodity approach is pushed too far. Just as a chess player who looks at his pieces one by one and is determined to preserve them all deprives himself of the use of the calculated sacrifice that so often brings worthwhile benefit to his whole situation, so a too zealous pursuit of commodity policies could block the way to more beneficial approaches with a broader aim in view. In any case, there must be some reforms that relate to the farm rather than to the marketing of produce and these could never be tackled by the commodity approach. For instance, the Country Landowners' Association would like to see the Community's aid to capital investment transmitted by means of capital grants instead of subsidised credit, as being a more efficient system.

The need to transform the CAP

With regard to the long-term objectives, whenever Britain's ministers,

66

officials and agricultural leaders can lift their sights from the immediate problems that encumber their desks and give consideration to long-term aspirations, the perspective is bound to change. They then realise that in the short-run they may have to agree to live with price levels (absolute or relative) and other market arrangements that in the longer run they will desire to change. Whereas piecemeal reform will have its place, making improvements here and there as opportunity offers, bringing about by persuasion the removal or abandonment of imperfections, the time will come when the UK may have to propose changes so considerable in their extent that they will amount to transformation.

Such transformation may prove to be organisational rather than a change in the whole philosophy of the CAP. Anthony Sampson, in his book *The New Europeans*, has analysed the administration of the CAP in terms of a delicate balance between the Council of Ministers and the Commission. The dialogue between them, he says, 'represents in miniature the battleground between the conflicting forces of Europe, between internationalism and nationalism'. The balance between them has constantly shifted. 'There were successive crises between Commission and Council, but a breakdown was averted … they wore each other down with negotiating in "marathon" sessions … eventually making bargains from sheer exhaustion … . It was an exacting and nerve-racking system, but it worked.' Whether or not this is an accurate analysis, it reveals a situation which, to say the least, should not automatically be extended in perpetuity. When our representatives, at all levels, have become familiar with the present mechanisms and all their imperfections, and when they have had opportunities to see how the CAP is fashioned and which are the channels and instruments that can be used most effectively, some realignment of forces and Community institutions is bound to occur. Our farmers' leaders will find out by experience on what kind of issues they can exert influence effectively through COPA and on what kind of issues it will remain preferable to use national channels of access to ministers and to Community officials. Our parliamentarians will have to consider what advantages there might be in strengthening the European Parliament as a means of bringing an influence to bear at the international level which they find they cannot exert through the accountability of British ministers to the House of Commons.

We have yet to see in what ways the British government will give expression to the vision of Europe which, as the Prime Minister has said, must transcend nationality in future. One cannot help being impressed by the whole-hearted devotion to British interests that is evident in the attitude of Whitehall today in the approach of public servants to the intricacies of negotiation. There is, understandably in the present circumstances of bargaining for initial positions, little outward sign of the kind of magnanimity that inspired

Churchill's European speeches. Yet eventually this other aspect of Britain's becoming part of the European Community will show itself more forcefully and her various agricultural institutions need to stand in readiness for that change of approach when she begins to shed 'the mystique of her own uniqueness'.

Exactly in which areas of agricultural policy such a shift of perspective will first make itself felt is hard to say, but in shaping the future attitude of the Community to the situation of the developing countries and the markets for their products, the British government will be anxious to exert its influence. If the CAP is acknowledged to be largely protective in character at this present stage of its history, there is need to re-examine the question – protective against whom, on what grounds, and at what cost to people within Europe and beyond its borders? Even as Britain moves forward from a predominantly nationalistic to a European approach to policy-making, the wider questions of regional disparities will thrust themselves more and more into view. It is to be hoped that her agricultural institutions, as they take up their new European responsibilities with vigour and resourcefulness, will not become imbued with a regionalistic devotion that denies the logic and reality of the world agricultural situation.

6 Agricultural Co-operatives in an Enlarged EEC

Sir Roger FALK

Chairman, Central Council for Agricultural and Horticultural Co-operation

Confidence in the Common Market as an organisation that will work rests on the assurance that, underlying the various political, social and economic systems it contains, there is a universality of ideas and experience which makes it possible for the member states, new as well as old, to understand and collaborate with one another. One such 'international language' is that of agricultural co-operation and, as there will be occasion later to dwell on the divergences in its development, it is right to begin by emphasising its basic unity. This unity is only in part historical. It is due mainly to the similarity of the situations in which primary producers in different countries find themselves; that is to say, with the common need to set up an economic service that will cater effectively for their commercial requirements as buyers of farm inputs and sellers of farm products, while avoiding the risk of domination either by outsiders or by one of their own number. These, not any ideological considerations, are the factors that induce farmers to establish co-operatives and determine the form they should take; they are factors that do not vary very much from one European – or, for that matter, non-European – country to another.

The task of preparing this paper has been made easier by three Central Council publications. The first, dealing with agricultural co-operative organisations in the EEC,[1] was published in 1970. This was followed in 1971 by a similar study of the UK,[2] and, in February 1972, by a publication dealing with Denmark, The Irish Republic and Norway.[3] This third study was made in the expectation that the three countries concerned would take up membership of the Community with the UK at the beginning of 1973.[4] All three studies consist of a descriptive account of the agricultural co-operatives and provide a good starting point for analysing their performance.

Assessment of agricultural co-operatives as an economic force

The figures given in Table 6.1 should be regarded as indications of magnitude rather than precise statistics. Collected from different sources at different dates and based on a variety of criteria, they purport to show what proportion of the national business in various agricultural fields is carried on by organi-

Table 6.1

Proportion of National Trade in Various Commodities Carried out by Co-operatives

	Belgium	Denmark	France	Germany	Ireland	Italy	Lux'bg	N'lands	Norway	UK
	(a)	(b)	(a)	(a)	(b)	(a)	(a)	(a)	(b)	(c)
					Percentages					
Cereals	20		82	45		15	70	55		6
Milk	50	85	52	83	80	29	91	84	100	6
Potatoes			15	22		32	50	14	100	
Meat	5[1]	92[1]	10	25	30[1]		26	20		
Fruit	60	100	25	22	30	10[2]	100	95	40[2]	
Vegetables	40	100	20	32	30	15		99		
Eggs	2	46		20				21		
Wine			37	30		6	70			15
	(b)	(b)	(d)	(d)	(b)	(d)		(d)	(b)	(c)
Feeding stuffs (milled)	50	50	27	33	20	20		48		
Fertiliser (sold)	40	40	50	60	40	55		60	62	18

[1] Pig meat
[2] Fruit and vegetables

Sources: (a) COPA-COCEGA, figures quoted in April 1971.
(b) CCAHC, *Agricultural Co-operative Organisations in Denmark, the Irish Republic and Norway* (1972).
(c) CCAHC, *Agricultural Co-operative and Related Organisations in the UK* (1970).
(d) UNCAA, figures quoted in February 1972.

sations which, in the countries of an enlarged EEC, are considered as being of a co-operative type.

These figures may convey a false impression. One has to remember that in each country the agricultural co-operative trade is being conducted by hundreds, if not thousands, of independent organisations, acting in an unco-ordinated manner and possibly in competition with one another. Secondly, their activity may take place at a low level; that is to say, while a producers' co-operative may be involved at the level of procurement or initial preparation for market, its throughput will often be taken over at the subsequent stage of processing or distribution by a non-co-operative organisation that could with equal justice claim the same produce as part of its market share. A similar situation is found in the supply field where, for example, a co-operative may be a manufacturer as well as a distributor of feeding stuffs, or, on the other hand, merely a distributor of feeding stuffs manufactured by others. The figures of market share may not therefore mean much in terms of market power. The extent to which they do mean something depends first on the nature of the market, which may itself be fragmented, in which case even a little marketing strength goes a long way, and secondly, on the extent to which the co-operatives have unified their forces and undertaken supplementary responsibilities; that is, the extent to which they have developed horizontal and vertical integration. To make an accurate appreciation of the significance of these percentages, it is necessary to inquire further into the extent to which commercial control has been centralised in second- or third-tier organisations and how far along the supply or marketing chain such organisations have been able to penetrate.

Before going further down this path of inquiry, it will be convenient to make a short digression to consider what encouragement agricultural co-operation has received from the member states of the Community and what has led them to give it. The main objective, it seems, has been to effect a structural improvement in order to make the farmers in a particular member state better able to compete with producers in other countries, so that the level of protection need not be so high, and that they may become less dependent on other forms of assistance. Capital subventions, reduced rates of interest and loan guarantees have all been used to this end, together with tax concessions. The means employed have not only been financial or fiscal. In the Netherlands, where it is often claimed that the co-operatives have had very little government assistance, for many years horticultural produce could be sold only through the auctions, which were co-operative. In Germany, a law of 1934 provided that all co-operatives must be members of a federation for audit purposes; in this way an important impulse was given to the setting up of central co-operatives for commercial as well as administrative purposes. In Norway a marketing act of 1930 imposed a levy on all producers, but only co-operatives can benefit from the fund thus created. The UK policy of encouragement to co-operatives is of recent origin and the type of assistance given at the present time is relatively modest, consisting of help with feasibility studies, formation costs and the appointment of key staff. Where capital aids are concerned, members of co-operatives are entitled to nothing more than that for which they would have been eligible as individuals. Since the per ton costs, of storage for example, fall as the size of the installation increases, the government could actually save money at the end of the day from this form of investment.

State aid to agricultural co-operatives is accorded mainly at the primary level, it being considered that if co-operatives cannot see for themselves the value of co-ordinating their efforts it is hardly the place of others to persuade them. Experience shows that co-operatives find difficulty in co-operating, for reasons that are more complex than they appear at first sight. Every agricultural co-operative has to reconcile within itself two different and, possibly, opposing points of view. On the one hand the farmer, that is to say the production manager, sees the co-operative as his organisation, which he owns, and in which he gives the orders. On the other hand, the chief executive of the co-operative knows full well that, whoever owns the organisation or gives the orders, it will succeed only if it follows the course dictated by the needs of the market. This problem is being encountered, and daily overcome, in agricultural co-operatives of all shapes and sizes. Generalising rather freely, one might say that the problem becomes somewhat easier to solve in organisations large enough for there to be some buffer of middle management between the level of production and the level of marketing, to enable the imperatives of the market to be translated into production terms, and vice versa. A direct confrontation between them may result in too great a tension. The tension is always there, however, and the larger the organisation becomes, the greater the risk that it may be regarded by the farmer as working in its own, and no longer in his own, interest. This suspicion is not always without foundation, though it is often exaggerated.

Agricultural co-operatives have a continuing need to explain themselves to their members in order that they may retain their confidence and be able to draw on them for commercial support and finance. This dependence partly explains the rather different development of agricultural co-operatives in the UK compared with other countries in Western Europe, where the existence of a degree of commercial monopoly, for example through the ownership of central grain storage, and the substantial amount of financial independence available through recourse to specialist institutions have enabled agricultural central co-operatives to forge ahead further and faster than would have been possible without such assistance. These continental co-operatives – for example, the French National Union of Agricultural

Cereals Co-operatives (UNCAC), the Dutch National Co-operative Purchasing and Selling Society for Farming and Horticulture, the German
Hauptgenossenschaften and the Federation of Danish Bacon Factories – are
in a position of considerable commercial power and, as the UK comes
closer to them, they will undoubtedly exercise an influence on the development of similar second-tier organisations in the UK. In fact the process
has already begun.

Before leaving the subject of agricultural co-operation as a commercial
force, one must refer to Eurograin, the international grain and oilseeds
brokerage third-tier organisation owned by the central co-operatives of
about a dozen Western European countries and controlled by them according
to their guaranteed use of it. This means, incidentally, that France and
Germany have the major say in its affairs. As may be imagined, the formation
of such a body was not an easy task and one would expect that if the need
should arise – as surely it will – for other co-operative undertakings on a
European scale, the co-operative centrals will try to make use of the existing
machinery rather than start all over again. Perhaps the first point of significance to be noted about Eurograin is that there was never any question of
confining its membership to members of the EEC. The project for forming
such a company was conceived within the European Confederation of
Agriculture, an organisation whose importance has been somewhat under-
valued in the UK because it has been concerned with words rather than
deeds. If the UK has not appreciated by now that international action takes
place only after a great deal of preliminary talking, she will certainly do so in
an enlarged EEC. Eurograin was launched in 1967. In order to have a stake
in the venture the leading British agricultural co-operatives formed a holding
company, Farmers Overseas Trading Ltd., to act as shareholder – the first,
but assuredly not the last, of the measures of national reorganisation to be
taken when Eurograin set up a British subsidiary, with FOT Ltd as its
commercial partner. As a brokerage concern, Eurograin has expanded
rapidly, inevitably doing a lot of business with both co-operatives and other
types of firm. In terms of economic power it still ranks a long way below the
international shippers of grain and oilseeds which, beside their trading
connections, hold important investments in shipping, port facilities and
transport.

The financial and legal situation of agricultural co-operatives

Co-operatives are not formed for reasons of investment, but rather to provide
those who set them up with a service which, in an agricultural co-operative,
is normally that of buying farm supplies more cheaply or of selling farm
products to better advantage. The aim is always to obtain these services

with as little investment of capital as possible, thus making more of it avail able for use on the farm itself. Early in their history, continental co-operative hit on a device that obviated the need for any capital investment whatever This was to set up organisations of unlimited liability, the operations o which were guaranteed and made credit worthy, first by the joint and severa guarantee of each of the members, and secondly by the undertaking of these members to use the co-operative to the exclusion of all other trading channels In the rural communities where this device was first applied and where members were close neighbours it worked extremely well. The result wa that very large organisations eventually grew up with what must now seen to be a ridiculously small amount of capital, if it is not appreciated that the sense of personal solidarity, even though never formally invoked, and perhap after so long an interval more than a little unreal, is still present in the relations between a European agricultural co-operative and its members. The British and Irish co-operatives have to be excepted from the European rule for their history began in quite a different way with a law of 1862 which imposed on co-operatives the then still fairly novel principle of limited liability, the aim being, needless to say, to protect the inexperienced co operators from the appalling failures that were such a feature of business life in nineteenth-century England and Ireland. So it becomes plain that, al though the principles of agricultural co-operation apply fairly uniformly through Western Europe, there can be important divergences of practice.

A second difference also has its roots in social history. In the mid-nine teenth century the banking system was already well established in rura England, and the English farmer, a person of some consequence as a resul of the enclosures of the Napoleonic Wars, had ready access to it. The situa tion over most of the Continent was basically dissimilar, with a need fo specialist credit institutions for peasant farmers. Following the lead given by Raiffeisen, many credit co-operatives were established. The lack of investmen in agricultural co-operatives was counterbalanced by savings in agricultura banks which, on the one hand, attracted the rural community in general a depositors and, on the other, found in the agricultural co-operatives a usefu outlet for their surplus funds. The result is that whereas in England farmer have usually been net lenders, as far as the banks are concerned, on the Continent there was, through the co-operative banks, a steady transfer o resources from the non-agricultural to the agricultural sector.

The consequence of these two points of difference between practice in the UK and on the mainland of Europe emerges very clearly in the balance sheets. A continental co-operative will be found in general to have an 'owned capital', consisting of shares, if any, reserves and undistributed profits constituting a very much smaller part of the total capital employed than it British counterpart; equally it will be found to have a very much larger par in the form of long-term loans, usually from a co-operative bank. To put this another way, if the gearing of owned capital to borrowed capital were the same in a British as in a continental co-operative, the British co-operative would be able to support much more, possibly twice the amount of borrowing, and a considerably greater turnover. This is something that has to be borne in mind very much when comparing the share of the market held by co-operatives.

Another interesting fact bearing on the same point is that continental co-operatives do not seem to make the clear-cut distinction that is made in the UK between 'owned' and 'borrowed' capital. Nor is there any reason why they should when so much of the latter consists of borrowings from banks which, though legally quite distinct from the organisations to which they make the loans are, as it were, members of the same co-operative family. This state of affairs, which is barely appreciated by UK co-operatives, is very much envied by them. In truth it may not be quite so enviable as it may appear; in recent years many of the continental co-operatives have begun to feel somewhat vulnerable, and one observes there the same desire to increase the proportion of capital that is held by the members individually in the form of shares, or corporately in the form of reserves, as is found in co-operatives in the UK. One of the reasons for this change of attitude is that the European agricultural co-operative banks or credit institutions, with their growing dependence on sources of capital outside agriculture, have to take a firmer line towards their co-operative creditors than they have adopted in the past.

Turning now from the financial to the legal situation, it is clear that an even greater diversity is to be found within the countries of an enlarged EEC, ranging from Denmark, which has no law for co-operatives, to France, which has several. On the whole, continental legislation is more precise than that of the UK, where considerable discretion is given to a government official, namely the Registrar of Friendly Societies, particularly in the important matter of deciding whether an organisation seeking to obtain incorporation under the Industrial and Provident Societies Act should be allowed to do so, and having become incorporated, whether it should remain on the register. (His writ does not run to co-operative-type companies, however.) There is nothing quite comparable to this in continental legislation. In addition, there are other important differences in the legislation of the existing EEC member states. Students of the subject have concluded that there can be no prospect of harmonising the laws under which agricultural co-operatives are recognised in the different countries of the Community, and perhaps not much practical necessity for doing so, but that there may be a case for promoting a supranational law for those co-operatives, like Eurograin for instance, that work internationally. In practice, the establishment of a European form of co-operative will take second place to the establishment of a European form of company, for which there is obviously much greater urgency, and it may be that if the latter is achieved the former will no longer be necessary.

Agricultural co-operation and the common agricultural policy

A supranational law would be an EEC law. Mention of this is a reminder that up to this point consideration has been given only to policies and practices of countries within the EEC or intending to join it, not as yet to those of the EEC itself. One should perhaps introduce this subject by recalling that the main instrument of the CAP is the European Agricultural Guidance and Guarantee Fund, and that of the two sections of this Fund, the guarantee section is at present by far the larger; the proportions may vary, though major changes do not seem very likely in the immediate future. Both sections are ultimately concerned with the structure of agriculture within the EEC, but the guidance section is concerned more directly. While it is a section whose policies and activities can be expected to be changed greatly as a result of the enlargement of the EEC, a start can be made by considering it as it is at present.

The structural problems of the EEC in relation to agriculture are necessarily the same as those of its members. Agricultural incomes are low by comparison with the incomes of those in other forms of employment. One of the ways of achieving higher agricultural incomes is to improve the bargaining position of producers in relation to other sectors of the food industry, that is manufacturers of farm inputs, processors, distributors and so on. The concept here is not that of a simple unification of producer effort – though this can be important too – but rather an improvement of effort by the production and marketing of goods in more desirable quantities and qualities. In order to achieve this result it is necessary for producers to group together; the question arises, therefore, of the form that these groupings should take. In view of the previous experience of the individual countries of the Community, one might have expected that a co-operative form of grouping would be encouraged. Such, however, was not the case. For reasons at which an outside observer can only guess, the Commission appears to have taken the view that the form of constitutional structure of such groupings is unimportant, so long as they have the correct objectives. Evidence of this policy is to be found in the EEC regulations that have appeared so far, or are proposed. A regulation of 1966, which was obviously to some extent experimental, provides for 'producers' organisations' in the field of fruit and vegetables; the function of these organisations is defined with some precision, but little or nothing is said about their internal arrangements. A 1971 regulation for hops and a draft regulation for 'producers' groups' over the whole field of agriculture are rather more specific, but still leave the matter of constitution exceedingly vague. There is nothing, for instance, to prevent such producer groups being dominated by one of their number, whose interest in production could be entirely secondary to that of wholesaling or processing. The proposal concerning the draft 'producers' groups' regulation has had a mixed reception, with the existing co-operative bodies arguing that it will invite proliferation and result in a weakening rather than a strengthening of producer interests. The debate recalls that which took place in the UK when the 1967 Agriculture Act was under discussion; no doubt, now as then, the points of view of both the innovators and the conservatives will be shown to have a certain amount of justification.

The discussion is one to which the UK ought to be able to make a useful contribution, in view of the fact that between 1967 and 1972 the problem of the relation of tightly knit groups of the kind envisaged in the EEC regulation and the more loosely organised agricultural co-operatives as they have developed in Western Europe has been explored in some detail. This contribution is likely to be made at two levels, that of member governments and that of the producers' own representative organisations. These are the Committee of Agricultural Professional Organisations (COPA) and General Committee of Agricultural Co-operation (COGECA), initials which one may expect to become as familiar as NFU and ACA (now ACMS) with which indeed they correspond at a European level. Having noted the existence of these channels of communication one need do no more than emphasise the importance of making good use of them. Our co-operative colleagues in Europe never fail to remind us that our representatives in this forum must be the best that the country is able to provide.

UK agricultural co-operatives in an enlarged EEC

The object of this paper is to examine agricultural co-operatives in an enlarged EEC in order to assess their effectiveness in any future European situation. But it must be apparent, even on the most summary acquaintance, that there is very little organisation of agricultural co-operation at an international level; indeed there is not much even at a national level. Most agricultural co-operatives are still firmly local in their outlook, even if the localities are somewhat wider than they used to be. Certainly this is the case so far as the UK is concerned. It would seem that this situation must change under the influence of the EEC, motivated by the CAP. There is in the larger French, German and Dutch agricultural co-operatives an awareness of the movement of events and the direction in which they are moving that as yet hardly exists in the UK. These organisations will respond to the situation and their response will be copied by others whose appreciation of the situation is at present less acute. Eurograin was formed in the first instance because the pace-setters realised the need for it, and it was soon extended when it became clear that the initiative was worth following up. It is probable that, given the anomalies that exist in the present arrangements for supplying the European agricultural market with various of its requirements, such as certain fertilisers and agricultural chemicals, other arrangements of a Eurograin type will be entered into. If this happens, and provided their commercial intelligence is good enough, the British co-operatives will want to follow suit, even if this involves the formation of national agencies that would otherwise never have entered into consideration. The important thing for the requisite British co-operatives at the present time is to become fully aware of the possibilities that will await them in the next few years; it is encouraging, therefore, that, at the present moment, the leaders of some of the most important of these co-operatives are trying, with the help of the Central Council, to arrange direct contacts with their opposite numbers across the Channel. Personal understandings are the essential foundation on which future commercial links must at all times be built.

Where the marketing of farm produce is concerned, it seems unlikely that there will be the same degree of international collaboration between UK and continental co-operatives. To the latter, the enlargement of the EEC means above all the opening up of the British market to their produce. Partly in order to be able to assess the strength of these co-operatives – though it may not be from this quarter that the strongest competition will come – the Central Council arranged during the spring and early summer of 1972 for leading British co-operative experts in nine different commodities (cereals, potatoes, beef, pigs, sheep, eggs, fruit, vegetables and salad crops) to spend a week investigating the marketing arrangements for these commodities on the other side of the Channel. One Central Council staff member was responsible for each visiting team, for making his separate evaluation and for co-ordinating the eventual reports. Without question this proved a valuable exercise, not least for showing that comparisons were not nearly as often to the disadvantage of the visiting team as it was thought they might be. These investigations overseas can be regarded as supplementary to the strategic studies in cereals, potatoes, eggs and vegetables that the Central Council has conducted, and is conducting, in collaboration with the farmers' unions and the central co-operative associations, the object being to work out and pursue agreed development policies instead of pulling in different directions, as has sometimes happened in the past. Following up these different leads involves a great deal of work, but no one minds that so long as it is productive; one can say that the results to date, though limited, have been satisfactory. One must make this general statement and leave the subject there, for it is apparent that each commodity has its own production and marketing techniques, and there is insufficient space to go into detail.

At this point some mention must be made, albeit briefly, of a form of agricultural co-operation that has to some extent been pioneered in the UK and is quite different in its character and in its objectives from the types mentioned hitherto. Co-operation in production has not, at the present stage, anything to offer on the European stage but, behind the scenes, there is possibly no form of development that is of greater importance. This is a bold claim; the grounds for it are that the production co-operative has immense cost-saving possibilities, either on existing investments or on a new investment, for those who take part in it. At this level co-operation has something of great value to contribute to the improvement of farm structure which is the UK's trump card on entry into Europe. Co-operation in production is of recent growth; it is a development with which the Central Council has been closely associated in the past and will, for the reasons just given continue to be associated in the future.

So far as the financial and legal situation of UK co-operatives is concerned, here too the Central Council has had a good look at EEC institutions, and has come to some conclusions. On the financial side there is undoubtedly a good deal to be learnt; the UK has been favourably impressed by the loan guarantee system operated in Belgium and the Netherlands, as being a cheap and effective method of extending credit, and the Treasury has been persuaded to give it a trial. A more difficult question is whether the possibility may exist for establishing any new rural credit institution that would be able to give the UK agricultural co-operatives the same sort of support as received by those on the Continent. For the present all our energies must be concentrated on the existing formulae – better understanding by producers of the need to support their co-operatives in proportion to their use of its services; good management to enable them to enjoy the results of that support; sound financial discipline to ensure that the co-operative is able to undertake the development that is essential to maintain an enthusiastic staff. There is also scope for some innovation, for example in forms of shareholding, in the creation of subsidiaries and in the establishment of revolving funds by means of which the members of the co-operative make a contribution to its capital each time they use the services and receive this contribution back after a given number of years. These matters are dealt with in the report of a working party on agricultural law, in which the unions and co-operative associations took part, under Central Council chairmanship. After two years of work on this subject, the Central Council has submitted its recommendations to ministers, once more with the backing of all the parties concerned. Only a few years back such unanimity would have been barely credible.

Finally, it will be useful to consider briefly the still embryonic institutions concerned with agricultural co-operation at the Community level, and try to estimate their importance and influence with the Commission and with the member states that have set them up. One of these institutions, whose future role is not yet clear, is a committee that has been set up fairly recently by all the central credit banks and credit institutions. For the moment the main central body concerned with agricultural co-operatives is COGECA, the representative body in Brussels, which has the opportunities for daily contacts with the Commission and can hardly fail to increase its influence there. Certainly it will not lack support from the central co-operative bodies in the member states as long as the Commission's policy towards co-operation remains as ambiguous as it appears to be at present. As is now generally known, the various agricultural co-operative bodies in the UK have recently resolved the problem of representation on COGECA, so that they now have a direct channel of communication to the Commission and to the other agricultural central co-operative associations in an enlarged EEC. Consultative arrangements between the UK bodies and COGECA began on 1 April 1972, so that a good understanding should have been reached by the time they took up full membership six months later. One could find no better illustration of the impact of an enlarged EEC on the domestic institutions of the UK, when even the prospect of joining it brings about changes that in previous times people have striven so hard, but in vain, to accomplish.

References

1) *Agricultural Co-operative Organisations in the EEC* (Central Council for Agricultural and Horticultural Co-operation, 1970).
2) *Agricultural Co-operative and Related Organisations in the UK* (Central Council for Agricultural and Horticultural Co-operation, 1971).
3) *Agricultural Co-operative Organisations in Denmark, the Irish Republic and Norway* (Central Council for Agricultural and Horticultural Co-operation, 1972).
4) This paper, which was finalised before the result of the Norwegian referendum was known, has made a similar assumption.

7   The Future Demand for Food in Western Europe

T. E. JOSLING
London School of Economics and Political Science

Food demand and farm policy

The theory of consumer demand is among the best established parts of economic theory. For any commodity that is purchased regularly and has a short life, purchases, and hence consumption, are determined by the interaction of the economic environment, that is the income of the family unit and the prices at which goods are available, and the social environment, usually summed up in the term 'tastes'. Economists have traditionally held 'tastes' to be relatively constant and hence concentrated on price and income as the main determinants of changes in consumption patterns. On the basis of fairly general assumptions about the consumers' ability and desire to choose among goods, it has been possible to suggest relationships between price and income and the quantity of particular goods purchased. For instance if the price of a product were to rise (and the consumer to be compensated for the effect on his total expenditure), then the theory would suggest a decrease in consumption. This 'law' of consumer demand seems to hold in practice. But the theory also leads to other predictions; for example, if all prices and money income rise by the same amount then consumption patterns will not change; if a commodity has close substitutes in consumption then it will tend to be price responsive in demand, and so on. With the help of these relationships derived from demand theory it is possible to build up a complete model of consumer demand for a range of different goods, and to estimate the parameters from past observation of market behaviour.

Although the theory of consumer demand is set at the individual or family level, it is common to assume that the same relationships will hold at the aggregate level. The estimation of demand relationships was among the first problems tackled by economists when the young science of econometrics emerged in the 1930s. Demand for various items of food was studied in detail, in particular by research workers in the Bureau of Agricultural Economics attached to the US Department of Agriculture. Since that time there must have been several hundred such studies in all parts of the world and at all levels of sophistication. In short, the economist should by now be able to estimate with reasonable accuracy future developments in the demand for food products in Western Europe.

Is there a need for such studies, beyond the natural curiosity of mankind about the future? Or should we look more closely at the demand conditions faced by groups within the agricultural sector? Certainly the overall level of consumer demand is relevant to retailing and transportation firms. It may be of interest to people who have to make investments in the food processing industry to know the extent to which, say, beef consumption will be contained by higher prices in the UK. On the other hand, total food consumption will be of less direct interest to the farmer. The demand curve faced by the individual producer can be taken to be perfectly price-elastic. He will be interested in future prices of both products and input items, but not directly in consumption levels. A group of producers, or a marketing organisation, will be concerned about the effect its own decisions will have on the market, but here again it is not so much the size of the total market as the conditions of competition with other groups at home and abroad that will determine its demand conditions. This applies particularly to the food industry, with a structure increasingly dominated by a few large firms. A cheese manufacturer, for example, will be more interested in the prospect of the exclusion of Australasian produce from the UK market than the change in the size of that market itself. Similarly, overseas suppliers will face a demand for their produce that is determined as much by domestic production levels as by total consumption patterns. The market prospects for any firm or group depend on much more than the total predicted size of the market in the UK or Western Europe.

One group of people has a direct interest in the aggregate demand for farm goods. Those who have to make policy decisions must be aware of the implications of their policies for consumers and users of agricultural products. Clearly food prices have become, and will remain, a politically sensitive issue and demand characteristics will help to determine the effect of a policy on food prices at the retail level. But more comprehensively, policy makers are in the position of acting as overall planners of the agricultural sector albeit without adequate control over production. A planner must know the parameters of the market. This element of planning has been somewhat obscured in the UK since the mid-1950s by the fact that imported produce in most major markets has been allowed in to satisfy demand over and above domestic production. Thus an accurate assessment of demand trends was not essential for meat and grain products; Annual Review calculations could be based on farm cost and income considerations. The market for the major commodities expanded fast enough to allow for the technical change in agriculture to be translated into higher production without severely restricting access to supplies from abroad. Those responsible for deciding on and administering the CAP were not allowed such innocence. The level of consumption has been important in EEC policy determination in a way that it has not been for the UK. Two clear examples are sugar and butter, where relatively small percentage changes in consumption show up as large changes in the activity of EEC intervention agencies and in the costs of export restitutions. The adoption of common financing of market support has meant that FEOGA costs cause political tension. This tension will be greatly increased on enlargement, when an ailing sterling has to support payments to producers in other countries for producing unwanted goods. Food demand influences policy in that it helps to determine the financial and political costs of policies designed to improve farm income. Policy-makers as planners must have an accurate picture of what will be the demand for food in the future.

More obvious is the influence of policy on food demand and on the demand for farm products used as intermediate inputs. Unfortunately most of these effects are quite unintended and would be considered undesirable in themselves. The Danish government has allowed a policy of high grain prices in an attempt to restrict pigmeat output, with the monopoly gains thus obtained in selling bacon in the UK market being passed on to grain growers. But this must be a rare exception. No one really benefits from a reduced consumption of, say, red meat. Nor are the potential gains to be made by companies that produce a substitute for high-priced butter a conscious objective of the EEC dairy regulations. There is much talk at present of introducing some form of income subsidy in the EEC to take the pressure off price supports. In so far as this allowed price rises to be moderated it would help to expand consumption, but a number of other policies could be developed with the same aim. Substitution of guaranteed minimum prices for butter in place of intervention buying would allow that product to regain the market lost to margarine at a considerable saving in cost. If a part of farmers' returns from selling grain were to come from a marketing certificate – somewhat along the lines of the scheme in use in the USA to urge compliance with acreage restriction – the price of feeding stuffs to the livestock farmer could be controlled. Expanding consumption appears to be the one policy direction that holds out hope both for a containment of costs within the EEC and also for a restoration of order in the international market. The wilful restriction of consumption through the use of inappropriate policy measures is to a large extent responsible for the present problems of the CAP.

Assumptions necessary for predictions of food demand

The pattern of food consumption is strongly linked to the level of available income and the prices prevailing in the market. As a consequence, the accuracy of predictions of future consumption is likely to be limited by the appropriateness of the assumptions made about the behaviour of these variables. Population projections are reasonably reliable and do not appear to cause problems for demand estimation. FAO, when recently checking the accuracy of their previous commodity projections, found that population growth rates had been very close to those used in their calculations. The same cannot be said for assumptions about income trends. Growth rates in many countries show somewhat surprising differences from one decade to the next. But even with income it is possible to use an estimate of the underlying growth in productive potential as a guide to future income levels. What has eluded economists to date is how to relate changes in government policy to rates of growth, if indeed such a connection exists. In the present context, for example, it has not been possible to get any consensus among economists who have studied the question as to whether entry into the EEC will affect the growth rate in the UK. Clearly this is as important in the food sector as in any other part of the economy. It has been estimated that an average rate of growth in consumer expenditure one half of one percentage point higher as a result of EEC entry would suffice to offset all the effect of higher food prices on consumption, save in the market for butter, over a period of ten years.

Price projections raise even more problems. Community policy operates in general at the wholesale or import level for raw or processed farm goods. Market conditions in the Community will determine the price level within the band of 'community preference', that is between the landed price of goods imported from third countries and the intervention floor of other members. When projecting prices it seems reasonable to take threshold levels for those where imports are expected, but this involves an interdependence as the level of self-sufficiency will itself depend on price. Ideally one should use a 'market clearing' model to determine consumption patterns where the price is likely to be between threshold and intervention levels. Clearly one must also have some indication of future prices to be set under the CAP. These will be influenced by the market as indicated earlier; where deficits persist there will be pressure to raise prices (as with maize and beef), while surplus commodities (wheat, butter, sugar, pig meat, apples) may experience slower price increases. Unless the system of support is modified there will be a strong tendency to align prices to desired levels of farm incomes. One can hazard a reasonable guess, therefore, as to the development of policy prices over the next decade, and by implication the price changes necessary for the new member countries to harmonise over the transition period. In view of the deep concern over the level of community prices among oversea suppliers one would expect a rather lower increase in cereal and dairy price than would otherwise be the case. A major fall in the world price for a commodity exported from the EEC would also tend to reduce price levels within Europe.

Consumers react to price levels at the retail level and respond to changes in prices expressed in their local currency. Exchange rate changes will affect trade flows and consumption patterns; parity changes will alter the domestic equivalent of the institutional prices unless offset by border tax adjustments. Such adjustments will be necessary so long as countries defend inappropriate parities by large-scale intervention in foreign exchange markets. There is no need, however, for such adjustments if parity changes are anticipated in setting future prices. Regular small parity changes are also, paradoxically, the only way that the infant European monetary union can be saved from the pressures imposed by national monetary autonomy. One would expect, therefore, to see prices in the UK rise, in domestic units, one to two per cent faster than those in Germany and the Netherlands.

The common system of indirect taxation, the value-added tax, is a tax on consumption collected, for administrative purposes, at several stages of production and distribution. At present the UK government intends to zero-rate food goods when the tax is introduced in 1973. But in the Six food is not zero-rated. As shown below in Table 7.1 the rates vary from country to country.

Table 7.1

Rates of Value Added Tax in EEC
member states

| Country | per cent | |
	Standard Rate	Food
France	17.6 and 23.0	7.5
Belgium	6.0	6.0
Netherlands	14.0	4.0
Luxembourg	10.0	10.0
Germany	11.0	5.5

Italy has yet to introduce a VAT system. There will be pressure on the UK both to harmonise its rate (at present expected to be 10 per cent) with other countries, which may entail some increase, and also to harmonise the base on which the tax is levied by including food. Presumably there will be border tax adjustments on food goods until the rate and coverage is uniform.

Common wholesale prices do not, of course, imply common retail prices.

The discrepancy between prices within the Six at present is striking. Processing, transportation and retailing costs are by no means uniform. In projecting retail prices in the UK it is better to start from expected changes at the farm-gate or wholesale level. One would expect marketing margins to increase with labour costs, but this rise will be modified by increases in efficiency. There is no reason to expect either constant percentage mark-ups over the cost of raw materials or, for that matter, constant margins per unit of product. The marketing margin is determined by costs in the marketing process and by the degree of competition faced by the firms in the industry.

These comments by no means exhaust the problems encountered in estimating demand. In particular two other aspects, namely technology and tastes, require examination. In so far as most farm goods undergo a degree of processing, demand will be related to the technology involved. Where agricultural production is used as an input item elsewhere in the farm sector, changes in both the type and efficiency of the transformation process will affect demand. But the choice of techniques is itself related to prices and profitability. Development economists have recorded the phenomenon of 'induced innovation' whereby production methods themselves are influenced by the relative prices of inputs and output. The Agricultural Adjustment Unit's work on farming systems would suggest that one can expect the same process to operate in UK agriculture. Specifically, the demand for feed grains could be reduced considerably by a switch to less intensive systems of livestock production. There is also the possibility that consumer tastes are influenced over time by price relativities. Items become established in diets, like poultry meat as an alternative to red meat. Other goods may come to be regarded as a luxury – there is the possibility that butter may suffer this fate. The implication of taste and technology change is that long-run demand elasticities may be much larger than those measured by observing the effects of year-by-year fluctuations in food prices.

An estimate of the demand for food products in Western Europe

The preceding section indicated the need for caution in the interpretation of estimates of future demand. The results presented in this section should be read in that light. They are drawn from a larger study designed to explore food import patterns in Western Europe.[1] The implications of the results of the Michigan State University study for the market for farm produce in the EEC are discussed in a paper by myself and Denis Lucey.[2] The study as a whole developed econometric models of the supply of and the demand for the major products in the grain-livestock economy of Denmark, Ireland and the UK. On the basis of assumptions about price levels over the decade the production and consumption patterns were projected. These results were then

86

co-ordinated with some updated estimates of the market balance in the Six obtained from an earlier study at Michigan State.[3] Only the demand estimates are presented here. The technique used in the demand analysis was common to the three applicant countries, and is described briefly in an appendix. It enabled consistent estimates to be made of the consumption of a number of commodities, including the interaction among the markets for related foods.

The necessary assumptions about the level of retail prices were obtained by forming an opinion as to the development of farm prices over the next decade, and adjusting for margins and processing costs.[4] The farm prices used in the projections are shown in Table 7.2. Prices for all commodities were expected to rise, in part due to the adoption of the CAP, and in part from general cost trends. In addition to the price assumptions it was also necessary to project income, population growth and inflation. Population trends were taken from official sources; all countries, including Ireland, were expected to increase their population at a low but steady rate. National income was projected on the basis of underlying growth in productive potential;

Table 7.2

Projected Indices of Producer Prices for Major Agricultural Products, 1980, Under Assumption of EEC Enlargement
1968=100

	United Kingdom	Ireland	Denmark
Barley	156	164	165
Milk	128	199	170
Fat cattle	170	168	216
Pigs	142	140	140
Lambs	140	154	–
Broilers	125	–	131
Eggs	107	92	104

the rates were 3.0 per cent for Denmark, 2.9 per cent for the UK and 4.2 per cent for Ireland. The study assumed a somewhat optimistic rate of price inflation of 4.0 per cent; although high by historical standards, this is probably not attainable for the decade in view of recent inflationary trends. For this reason, the retail prices are probably understated, but with little effect on the consumption pattern.

The direct price and income elasticities at retail for the major farm commodities are shown in Table 7.3. The data required for the estimation of these

Table 7.3

Price and Income Elasticities, for Food Demand at Retail, 1968, UK, Ireland and Denmark, as Used in Demand Projections

Commodity	United Kingdom		Ireland		Denmark	
	elasticity with respect to own price	income	elasticity with respect to own price	income	elasticity with respect to own price	income
Beef	−2.49	0.71	−0.10	0.45	−0.01	0.27
Pigmeat	−2.37	0.61	−1.27	0.24	−1.37	0.26
Mutton and Lamb	−1.35	−0.10	−2.55	0.75	–	–
Poultry	−0.25	0.79	−0.01	1.23	−0.27	0.40
Milk (liquid)	(a)	(a)	−0.05	0.10	−0.32	0.13
Butter	−0.38	0.60	−0.02	0.56	−0.10	0.57
Cheese	−0.12	0.39	−0.01	0.79	−0.13	0.90
Margarine	−0.28	−0.49	−0.07	0.15	−0.29	−0.44
Wheat Flour	−1.02	−0.48	(a)	(a)	−0.01	−0.16
Bread	0.31	−0.76	−0.15	−0.17	(a)	(a)
Eggs	−0.16	0.18	−0.09	−0.51	−0.17	0.26

[a] Projections of trends used in place of demand equations.

parameters came from published retail price and consumption statistics. For the UK use was made of the comprehensive data available in the National Food Survey published annually by the Ministry of Agriculture, Fisheries and Food. The projections were adjusted to include food consumption not covered by the NFS, such as institutional demand and meals eaten outside the home, so that total consumption corresponded to the domestic disappearance figures published by the Board of Trade. In the case of Ireland, the consumption and price data came from the Irish Statistical Bulletin; annual prices were calculated from the quarterly figures as an unweighted average. For Denmark, use was made of a report from the Landboforeninger of consumption and prices; use was also made of the study by Professor Vibe-Pedersen and his colleagues at Aarhus University, which covered much the same ground.[5] The demand for grains was estimated partly from the retail consumption of bread, wheatflour and other grain products and partly from an examination of the needs of the livestock industry for concentrate feed derived from the production estimates contained in the study and based on an assumption of stable feed composition.

The projected consumption of the major food products in 1980 is given in Table 7.4. Of the meats, consumption of beef is expected to decline by 6 per cent in the UK and to rise in Ireland and Denmark by 20 and 15 per cent respectively. But demand is expected to continue strong in the Six, and taking the nine countries as a whole beef consumption rises by over one-quarter relative to 1968. The decline in the UK reflects the effect on demand of a substantial price rise over the period, more than offsetting the increased demand through income and population growth. Meats that are quite close substitutes for beef benefit from this relative price effect. The pig meat and the mutton and lamb markets experience growth; the demand for pig meat throughout the enlarged Community is expected to rise from just over 6 million tons to nearly 8 million tons by 1980. Poultry meat consumption is also expected to increase, but even at these higher levels, *per capita* consumption of poultry meat is small compared with the US.

The demand for milk and milk products is less buoyant. In Table 7.4 the total demand is expressed in butterfat equivalent. Within that total the consumption of liquid milk shows little change. Retail demand for butter, however, appears to be quite sensitive to price changes. The high levy facing third-country suppliers and the internal market support system of the CAP depresses consumption of butter in the UK to about 85 per cent of 1968 levels. Consumption of cheese is also depressed by the imposition of levies in the UK. The total demand for milk and milk products in the enlarged EEC shows a small rise by 1980. For eggs a similar small rise in demand is expected. Demand for grains as a whole is expected to grow over the decade, due mainly to an increase in livestock production. Demand for food grains

Table 7.4

Consumption of Major Farm Commodities, 1968 and Projected 1980,
Assuming Enlargement of the EEC and Continuation of the Present CAP

'000 metric tons

Commodity	Denmark		Ireland		United Kingdom		EEC (six countries)		Total (nine countries)	
	1968	1980	1968	1980	1968	1980	1968	1980	1968	1980
Grains	6,132	8.786	1,380	1,639	21,724	22,989	73,271	87,557	102,507	120,971
Milk (in butterfat equivalent)	113	129	72	83	1,030	1,248	2,094	2,574	3,309	4,034
Beef and veal	92	106	52	62	1,130	1,063	4,341	6,001	5,615	7,232
Mutton and lamb	(a)	(a)	32	54	582	738	(a)	(a)	(a)	(a)
Pig meat	159	227	73	88	1,216	1,470	4,717	6,057	6,165	7,842
Poultry meat	24	31	25	47	509	688	1,744	2,898	2,302	3,664
Eggs	57	66	39	31	855	1,008	2,264	2,955	3,215	4,060

a not estimated

is virtually static, the growth in population just offsetting the decline in the market as consumers in their affluence eat less bread.

It is not easy to compare these results directly with those from other studies. Projection periods and price assumptions vary, and the products covered do not always coincide. The steady growth in demand for livestock products, and in particular for meats, is in line with other estimates. Similarly, other studies have suggested a slow rise in the demand for grains. Perhaps the most contentious results are those that indicate substantial substitution among commodities in the UK consumption pattern, particularly the decrease in the demand for beef and for butter. Other studies have either excluded such substitution by ignoring relative price changes[6] or have used an economic model where consumers are assumed not to react to these changes.[7] But the reaction of British consumers to recent price rises for beef and butter seems to confirm that they are very conscious of prices and adjust buying habits accordingly. One area of product substitution that may be understated in the results described above is that which might occur in the market for animal feed. In other countries feed compounders have reacted quickly to the higher prices for cereals implied by the CAP regulations.[8] There is a strong possibility that UK feed firms will react similarly. Sturgess and Reeves have calculated that even a modest degree of substitution could reduce total usage of grain to about 18 million tons by 1978.[9] This would reduce the size of the EEC grain market by a further 3 million tons, and could even lead to self-sufficiency by the year 1980.

Interpretation of food demand projections

These results to not represent a forecast. They are a 'best estimate' of the implications of a set of price assumptions based on the analysis of observed behaviour of consumers in the past. For this reason the results are of more direct interest to governments that to private firms. They indicate the pressures that will build up if present price and policy trends are maintained. It would be naive to expect policy-makers not to try to anticipate problems; these projections are intended to give an indication of the size of these problems in store for the politicians and civil servants of the EEC.

The most significant implications are connected with consumption changes, in particular product substitution in the UK. These are:

(a) a cutback in cereal consumption that will mean little if any outlet in the UK for the EEC wheat surplus. This will continue so long as the users of grains face such high relative prices;

(b) a decrease in butter consumption that will again solve none of the problems of market balance in the dairy sector, and, moreover, will

91

make it difficult to make provision for imports from New Zealand and Australia;

(c) the fall in the beef market will create trading problems with South America and Australia; although mutton consumption is not reduced, any attempt to introduce high variable levies on sheep meat would have serious consequences for UK consumers;

(d) the demand changes go far towards offsetting the balance of payments problems that the UK would face on return to a fixed parity. The exact extent of this offset depends on supply response as well;

(e) external pressures from overseas suppliers will mount as the cumulative effect of the UK import pattern is felt. The EEC will be under stronger pressure to react positively to proposals about limiting the trade-distorting effect of the CAP.

In the face of these changes and pressures it seems likely that within the next five years the CAP will be modified so as to ameliorate the worst of the consumption effects. The desire to control inflation and to preserve world trade will probably force a development of the CAP in the direction of divorcing producer returns from user prices. A realistic forecast of the market situation in 1980 might with justification be more optimistic.

References

1) Ferris, J. N., *et al.*, *The Impact on US Agricultural Trade of the Accession of the UK, Denmark, Ireland and Norway to the EEC*, Research Report No. 11, Institute of International Agriculture (Michigan State University, East Lansing, 1971).

2) Josling, Tim and Lucey, Denis, 'The Market for Agricultural Goods in an Enlarged Economic Community', a paper given to the Irish Agricultural Economics Society (Dublin, October 1971).

3) Production and consumption figures for Norway, at that time an applicant country, were obtained from OECD estimates. These figures have been removed from the tables in the present paper.

4) The next section leans heavily on the description in Josling and Lucey, *op. cit.*

5) Anderson, P. S., *et al.*, *Projections of Supply and Demand for Agricultural Products in Denmark* (Aarhus Universities Økonomiske Institut, Aarhus, 1969).

6) For instance the FAO projections are made at 'constant prices' and assume no change in farm policies. See *Agricultural Commodity Projections, 1970–1980*, CCP 71/20, (FAO, Rome, 1971).

7) See, for example, McFarquhar, A. and Hannah, A. C., 'Trade and the Changing Structure of Food Demand', in Josling, T., *et al. Burdens and Benefits of Farm Support Policies* (Trade Policy Research Centre, London, 1972).

8) See Pearson, W. and Friend, R., *The Netherlands Mixed Feed Industry – Its Impact on Use of Grain for Feed*, ERS-Foreign 287 (USDA, May 1970).

9) See Sturgess, I. M. and Reeves, R., *The Potential Market for British Cereals*, (Home-Grown Cereals Authority, London, 1972).

Appendix

Method for estimating future demand for food at the retail level

The general procedure was to establish demand relationships linking *per capita* consumption of each food good with its own retail price, the price of each of the other food goods, non-food prices, and money income level. The prices of all goods were then specified along with money income for the period 1968–80 under various policy 'case' assumptions. Prices were not estimated by the demand model; they were established in a separate routine in the computer programme from assumptions about farm prices and about the behaviour over time of retail-farm margins.

It was decided to work with proportional changes in price and quantity variables. Each demand relationship comprised a set of elasticities. These were allowed to change from year to year, except where constant elasticities were imposed after examination of past data. For each commodity the relationship between *per capita* consumption in one year and the value in the next was:

$$\frac{D(I)_t - D(I)_{t-1}}{D(I)_{t-1}} = \sum_{J=1}^{N} E(I, J)_{t-1} U(J)_t + E(I, Y)_{t-1} U(Y)_t$$

$$D(I)_t = D(I)_{t-1} \left[1 + \sum_{J=1}^{N} E(I, J)_{t-1} U(J)_t + E(I, Y)_{t-1} U(Y)_t \right]$$

where $D(I)_t$ refers to the *per caput* consumption of good I in time period t; $E(I, J)$ refers to the elasticity of the quantity of good I and the price of good J ($J=I$ for the direct price elasticity, and $J=N$ for the elasticity of consumption with respect to non food prices); $E(I, Y)$ is the income elasticity of good I; $U(J)_t$, $U(Y)_t$ are the proportionate changes in prices and income from year $t-1$ to year t.

The elasticity values were computed in three ways: (a) time series regression analysis; (b) implicit cross elasticities from budget constraints; and (c) implied cross elasticities with non-food prices from an assumption of zero-degree homogeneity. For each commodity, regression analysis on the quantities and non-deflated prices and income were used to derive the elasticities of the consumption of each good with respect to its own price, the price of *a priori* substitutes and complements (that is those goods for which it was expected in advance that the cross elasticity would be either positive or negative), and with income. For those pairs of goods where no a priori relationship was established, the cross elasticity was derived from the implicit effect on expenditure of a price change. If the price of good I rises by $U(I)$ per cent, then expenditure on that good increases by $[1+E(I, I)]. U(I)$ per cent, where

$E(I, I)$ is the direct price elasticity. The effect on total expenditure is therefore

$$A(I)[1+E(I, I)] \cdot U(I)$$

where $A(I)$ is the proportion of total expenditure accounted for by that commodity. The effect on consumption of good J is thus:

$$-E(J, Y) \cdot A(I)[1+E(I, I)] \cdot U(I)$$

and the cross elasticity between goods J and I is

$$E(J, I) = -E(J, YA) \cdot A(I)[1+E(I, I)].$$

This relationship only holds where $A(I)$ is small, so that there is no appreciable effect on the marginal utility of money arising from the change in I's price. The change in non-food prices could not be handled this way; instead it was decided after examination of other methods to derive the cross elasticities of food demand with non-food prices by restricting each food demand relationship to be homogeneous of degree zero. This implies that general inflation throughout the economy does not change the relationship between the quantities of the various food goods. 'Money illusion' is absent from food purchases; in other words a 10 per cent change in all prices and money income leaves the consumption pattern unchanged. This meant that the cross elasticity for good I with respect to non-food goods was

$$E(I, N) = -\sum_{J=1}^{N-1} (I, J) - E(I, Y).$$

In this way all the elasticities and cross elasticities were established and used to compute demand changes corresponding to assumed price and income shifts.

8 Supplies, Incomes and Structural Change in UK Agriculture

B. H. DAVEY
Agricultural Adjustment Unit,
University of Newcastle-upon-Tyne

Introduction

This paper is based on the preliminary results of a research project currently under way in the Agricultural Adjustment Unit. The principal objective of the research is to construct a supply model capable of predicting the aggregate supply of the major agricultural commodities in the UK under alternative policy assumptions. As such, it is an extension of earlier work undertaken for the Unit by Barnard, Casey and Davey[1] and Davey and Weightman[2].

The basic methodological approach is that of programming a set of representative farms that have been selected to represent groups of farms of different types and sizes in different parts of the country. The approach is thus a micro-economic one, containing five basic steps;

(a) To stratify all farms in the population into homogeneous groups;
(b) to define a representative farm for each group;
(c) to derive supply functions for each farm using linear programming;
(d) to aggregate the results of each representative farm according to the number of farms in the population which it represents;
(e) to adjust the results to take account of aggregate constraints not built into the individual representative farm models.

Basic assumptions

Underlying the projections of supply are a variety of assumptions on farm prices and costs, technical performance, gross margins and so on. These assumptions are described, and the rationale underlying them explained, in the following paragraphs.

Farm prices

Following the price increases agreed for the 1972/3 season, little general increase in EEC farm prices can be expected over the transition period to 1978. It seems unlikely that in the enlarged Community farmer pressure and cost inflation will lead to EEC prices rising in money terms on average by more than 1 per cent per annum. This is because any change that widened the gap between EEC and 'world' prices would stimulate European production and increase the cost of financing the CAP. A further constraint is the interrelationship between the CAP and world agricultural trade; this is, of course, a sensitive question at the present time and any proposals to raise EEC prices by more than a marginal amount is likely to lead to strong protests – and possible retaliation – from the traditional, low-cost agricultural exporting countries of the new world.

While the general price level may show little change between now and 1978, it is likely that price relativities between different commodities may alter in an attempt to influence the pattern of production. Information on UK and EEC official farm prices for the 1972/3 farm year is given in Table 8.1, together with estimates of prices payable to British farmers in 1978/9. A comparison of UK prices in 1972/3 with the Newcastle estimates of prices in 1978/9 indicates that substantial increases in farm-gate prices of cereals, beef, pigs and milk are in prospect, with smaller increases for eggs, poultry meat and sugar beet. Little change in the price of potatoes is expected, but while total returns from sheep meat will remain fairly stable, market prices will rise substantially following the phasing out of the deficiency payment scheme and introduction over the transition period of the EEC's common external tariff on mutton and lamb imports.

Two factors underly the predicted increase in milk prices; an appreciable increase in the return from milk sold for manufacturing following the introduction of EEC threshold price arrangements for imports of dairy products; and the expectation that the liquid milk price will be determined primarily by the price at which milk can be imported. So far as beef prices are concerned, a further stimulus to beef production can be expected over the next few years in an attempt to modify the balance between milk and beef production in favour of the latter. Higher beef prices, in turn, will stimulate an increase in demand and higher market prices for other meats and eggs through the substitution effect.

On the crops side, the projected increases in grain prices reflect the general arguments outlined above. There may be some scope for raising sugar-beet prices following the reallocation of at least part of the Australian quota under the Commonwealth Sugar Agreement to European beet producers. But little change in potato prices is expected, with the British producer effectively shielded from European competition in normal seasons by the effect of transport costs.

Costs

As well as receiving higher prices for the products they sell, farmers will pay

Table 8.1

EEC Target and Intervention Prices and UK Guaranteed Prices, 1972/3;
Estimates of UK Producer Prices, 1978/9*

Commodity	Type of price	Unit	Price
Soft wheat	EEC target, 1972/3	£ per ton	47.41
	EEC basic intervention 1972/3	£ per ton	43.65
	UK guarantee, 1972/3	£ per ton	34.40
	UK producer price, 1978/9	£ per ton	44.40
Barley	EEC target, 1972/3	£ per ton	43.44
	EEC basic intervention, 1972/3	£ per ton	39.88
	UK guarantee, 1972/3	£ per ton	31.20
	UK producer price, 1978/9	£ per ton	41.40
Potatoes	UK guarantee, 1972/3	£ per ton	16.55
	UK producer price, 1978/9	£ per ton	16.65
Sugar beet	EEC producer minimum, 1972/3	£ per ton	7.37
	UK guarantee, delivered factory, 1972/3	£ per ton	8.00
	UK producer price, farm-gate 1978/9	£ per ton	8.40
Fat cattle	EEC guide price, 1972/3	£/live cwt	15.63
	UK guarantee, 1972/3	£/live cwt	13.20
	UK producer price, 1978/9	£/live cwt	19.20
Pig meat	EEC basic price, 1972/3	£/score dw	3.10
	UK guarantee, 1972/3	£/score dw	2.81
	UK producer price, 1978/9	£/score dw	3.33
Eggs	EEC m.i.p. (Oct./Dec. 1971)	p per dozen	19.1
	UK guarantee, 1972/3	p per dozen	16.0
	UK producer price, 1978/9	p per dozen	18.6
Milk	EEC target, 1972/3	p per gallon	23.0
	UK pool price, 1972/3	p per gallon	20.6
	UK producer price, 1978/9	p per gallon	24.5

(continued) Table 8.1

Commodity	Type of price	Unit	Price
Broilers	UK producer price, 1972	p per lb	7.0
	UK producer price, 1978/9	p per lb	9.8
Sheep	UK guarantee, 1972/3	p per lb dw	24.3
	UK producer price, 1978/9	p per lb dw	24.3

* EEC official prices have been converted on the basis of £1 sterling equals 2.40 units of account, and thus take no account of the floating pound.
Source: UK prices 1972/3: *Annual Review and Determination of Guarantees 1972*, Cmnd 4928; EEC prices 1972/3: calculated from paper by J. van Lierde (chapter 4, Table 1). UK producer prices 1978/9: estimated by AAU.

more for the inputs they buy. In particular, higher ingredient prices will push up the price of compound animal feeding stuffs by 1978. Feed grain prices will rise by some 80 per cent in step with the transition from UK to EEC price levels and the change in price support methods. Prices of oil cakes and protein feeds are expected to rise by some 30 per cent, so as to bring forth the required supplies in the face of declining real prices for vegetable oils. Prices of cereal substitutes, such as grass meal, starchy roots, beans and cereal offals, will be bid up, perhaps by as much as 40 per cent, by increased demand, as these items are substituted for more expensive feed grains. With feed grain prices increasing relatively more than prices of other ingredients, changes in the composition of animal feedingstuffs are to be expected, with grains being replaced by other feeds along similar lines to developments in the Netherlands over the last decade. Nevertheless, substantial increases in ration costs are probable, with prices of typical compound cattle, pig, layer and broiler feeds reaching £52, £58, £58 and £69 per ton respectively by 1978 [3].

Removal of the fertiliser subsidy, already started in the 1972 Annual Review, under EEC rules governing fair competition between member states, will lead to increased fertiliser prices. Any further increases in price will be countered by the opening of the market and the excess capacity in the industry throughout Western Europe generally. As a consequence, it is expected that fertiliser prices will rise by only 2 per cent per annum over the transition period.

Increases in agriculture's fixed costs are also to be expected. Agricultural wage rates will be rising with inflation and economic growth. In its current work the AAU has assumed an annual rate of wage inflation in agriculture of 8 per cent, 4 per cent on account of inflation and 4 per cent representing higher real incomes for agricultural workers. For the other items – fuel, machinery, veterinary bills and medicines, minor farming requisites – an annual inflation of 6 per cent for the first two years of the supply forecast period and 4 per cent thereafter has been assumed; in the light of recent events this is, perhaps, too modest an assumption.

Technical change

Since the war there has been a marked improvement in the productivity of British agriculture. Average yields of both crop and livestock products have increased and there has been greater efficiency in the conversion of feeding stuffs by livestock into meat, milk and eggs. There is no reason to expect that this tendency will come to an end during the 1970s. Not only are agricultural scientists active in developing new technology for application to commercial farming practices, but considerable scope exists for an improvement in average performance through better management and a more widespread use of the best practices currently available.

Projected improvements in selected performance measures over the period 1969/70 to 1978/9 are summarised in Table 8.2.

Table 8.2
Selected performance Measures for UK Agriculture
in 1969/70 and Projections to 1978/9

Performance measure	Unit	1969–70	1978–9
Milk yield per cow	gallons	850	892
Lambing percentage – lowland flocks	lambs/ewe	141	151
Egg yield per hen	dozen eggs	18.9	21.3
Pigs reared per sow	pigs/annum	15.5	17.5
Sugar beet yield	tons/acre	15.5	17.5
Maincrop potato yield	tons/acre	10.5	12.0
Spring barley yield	cwt/acre	29	33
Winter wheat yield	cwt/acre	33	37.2

Source: calculated by AAU from past trends.
Improvements in feed conversion efficiency have also been assumed.

Gross margins

The assumptions on prices, costs and technical performance generate forecasts of gross margins; that is the margin of output over the variable costs of production. These forecasts are summarised in Table 8.3. It should be

Table 8.3

Estimates of Gross Margins for Selected Enterprises in 1969/70 and 1978/9

Enterprise	Unit	Gross margin per unit	
		1969/70	1978/9
Crops		£	£
Winter wheat	per acre	35	64
Spring barley	per acre	26	51
Spring oats	per acre	26	43
Potatoes (maincrop)	per acre	99	123
Sugar beet	per acre	65	88
Variable costs of:			
1 year ley (semi-intensive)	per acre	10	19
3 year ley (semi-intensive)	per acre	8	15
Kale	per acre	10	18
Beef cattle			
Intensive 12-month beef	per head	15	4
Semi-intensive 18-month beef	per head	51	68
Suckler herd (spring calving)	per cow	63	70
Dairying			
Concentrate-fed herds	per cow	95	147
Bulk-fed herds	per cow	101	154
Sheep			
Fat lamb production	per ewe	10.5	13.0
Upland ewes	per ewe	5.0	4.5
Mountain sheep	per ewe	3.5	2.5
Poultry			
Egg production	per hen	1.0	1.25
Broilers	per 10 birds	0.3	0.3
Pigs			
Weaner production	per sow	48	59
Porkers	per sow	88	101
Baconers	per sow	119	135
Heavy hogs	per sow	128	123

noted (a) that the margins for ruminant livestock are before deducting forage and grazing costs; and (b) that phasing out of the hill farming grants has been assumed in forecasting margins from upland and hill sheep and hill cattle.

On the basis of the earlier assumptions on prices and technical performance, it seems that there will be some increase in gross margin for most enterprises to offset or cushion the effect of higher fixed costs. But it is the manner in which the relativities between and within enterprises change and the way in which farmers will respond to these shifts in relative profitability that is the most interesting question raised by the figures in Table 8.3.

Some preliminary results of the Newcastle model of UK agriculture

While some forty individual farm models have been specified for England and Wales as a whole,[4] at the time of writing predictions are available for only a small number of farming situations. In this section of the paper the results for one from each of the main types in the Ministry of Agriculture's farm classification is discussed on the assumption that its behaviour will be typical of the whole group of these farms.

The objective of each model is to maximise a measure of profit equivalent to the return to the farmer's own labour, capital and managerial input. In specifying the individual representative farm models, a range of alternative activities was selected, comprising those enterprises produced in the base year, together with any enterprises that might reasonably be considered by the farmer in the future. In addition, a wide range of technological systems was specified within each enterprise. These production and technological alternatives were constrained by the usual resource, husbandry and capital considerations.

Dairy farms

Programming results for two groups, namely small, predominantly dairy farms in the Midlands and large, mainly dairy farms located in Wales, the North, South-West and West Midlands, are summarised in Table 8.4.

In both groups some expansion in the size of the dairy herd can be expected by 1978/9, with the average number of dairy cows increasing from 48 to 56 on the small dairy farm and from 106 to 142 on the larger farm. The sale of calves will increase, even after provision has been made for herd replacement. The expansion of cow numbers is accompanied by an improvement in the level of grassland management, particularly on the large, mainly dairy farm, where an expansion of 30 acres in the cereals acreage is also expected. Some contraction in the beef enterprise, particularly barley beef, and pig production will occur.

101

Table 8.4
Optimal Farm Organisation on Two Dairy Farms, 1969/70 and 1978/9

Type of farm Size of farm	Predominantly dairy Small		Mainly dairy Large	
	82 acres		372 acres	
Cropping (acres)	1969/70	1978/9	1969/70	1978/9
Winter wheat	1	1	89	105
Barley	11	11	89	105
Potatoes	5	5	4	6
Sugar beet	–	–	1	1
Kale	10	12	23	30
Stocking (numbers)				
Dairy cows, bulk fed, winter milk	48	56	106	142
18-month beef	1	1	8	8
Summer finishing, autumn born stores	–	–	1	1
Barley beef	–	–	4	–
Ewes	1	1	18	18
Winter fattening of store lambs	–	1	–	–
Breeding sows	1	–	3	–
Fattening bought in weaners to bacon	–	–	113	–

Livestock farms

Three farm models representing livestock types are given in Table 8.5. These are medium/large mainly cattle farms, small/medium mainly sheep farms and medium/large mixed cattle and sheep farms. On all three farms, the emphasis of adjustment is on cattle, particularly the size of the beef-breeding herd so far as the cattle and sheep and mainly cattle farms are concerned. In the base year 1969/70 both these farms had a herd of spring-calving beef cows producing suckler calves for sale as stores, which is projected to increase, especially on mainly cattle farms. In addition, an autumn-calving beef herd is introduced. As a consequence, the total beef-breeding herd on both farms is expected almost to double between 1969/70 and 1978/9. On the mainly cattle farm this expansion will be accommodated by the disappearance of a small dairying enterprise and a reduction in the acreage of potatoes, but more specifically by a more effective utilisation of grassland. Dairying also disappears on the cattle and sheep farm and so does barley beef, but in this case the main burden of facilitating an expansion in the beef herd falls on sheep. While it is unlikely in practice that sheep will be entirely eliminated from this type of farming system, greater emphasis on cattle and less on sheep can be expected in view of relative changes in the profitability of the two enterprises.

Table 8.5

Optimal Farm Organisation on Three Livestock Farms
1969/70 and 1978/9

Type of farm Size of farm	Cattle and sheep Medium/large 478 acres		Mainly cattle Medium/large 354 acres		Mainly sheep Small/medium 188 acres	
Cropping (acres)	1969/70	1978/9	1969/70	1978/9	1969/70	1978/9
Wheat	36	36	33	32	20	9
Barley	45	45	33	33	20	32
Potatoes	10	16	19	10	–	–
Sugar beet	–	–	1	1	–	–
Kale	–	–	1	–	1	–
Stocking (numbers)						
Dairy cows	2	–	5	–	2	–
18-month beef	28	23	22	22	5	13
Suckler cows, autumn-calving	–	69	–	33	–	–
Suckler cows, spring-calving	88	94	71	99	19	16
Barley beef	3	–	–	–	6	–
Ewes	266	–	–	–	290	290
Fattening purchased weaners to bacon	22	–	54	–	–	–
Breeding sows	–	–	–	–	1	20

No change is predicted in the size of the ewe flock on the sheep farm. Here again the emphasis in adjustment is on cattle, although some switch of resources between spring wheat and barley is also expected. Within the cattle enterprise the main changes are the disappearance of dairy cows and barley beef and a small contraction in the beef-breeding herd. At the same time, the size of the eighteen-month beef enterprise more than doubles.

Mixed farms

Only one mixed farming model is given in Table 8.6. This relates to small mixed farms of the type often found in Wales, North and South-West England.

Table 8.6
Optimal Farm Organisation on One Mixed Farm
1969/70 and 1978/9

Type of farm Size of farm	Mixed Small	
	78 acres	
Cropping (acres)	1969/70	1978/9
Winter wheat	3	3
Barley	12	12
Potatoes	12	11
Kale	3	4
Stocking (numbers)		
Dairy cows	12	17
18-month beef	14	17
Beef cows	12	19
Ewes	29	29
Fattening bought-in weaners to bacon weight	49	—

No change in cropping pattern is expected, except for a marginal diversion of resources from potatoes to kale to cater for the enlarged dairy herd. Once again the adjustment is centred on the cattle enterprise. Both the dairy and beef herds are projected to expand, with more emphasis on beef. The beef-breeding herd will increase by over 50 per cent and there will be a smaller expansion in the production of eighteen-month beef from dairy-type calves. No change is expected in the size of the ewe flock. The increase in grazing livestock will be accommodated by an improvement in the level of grassland management.

Pig and poultry farms

The programming results for a medium/large pig and poultry farm are summarised in Table 8.7. This model relates to specialist pig and poultry production in the Midlands and the North.

Table 8.7

Optimal Farm Organisation on One Pig and Poultry
Farm 1969/70 and 1978/9

Type of farm Size of farm	Pigs and poultry Medium/large	
	112 acres	
Cropping (acres)	1969/70	1978/9
Barley	63	63
Sugar beet	2	2
Potatoes	3	6
Kale	5	5
Stocking (numbers)		
Dairy cows	22	25
18-month beef	8	8
Barley beef	1	–
Store lamb production	17	–
Sows: for weaner production	2	1
for bacon production	54	16
Laying hens (10-bird units)	–	42

The basis of this type of farming in 1969/70 was a large barley enterprise with the grain produced fed to pigs. Apart from a small acreage of root crops the remaining land was used to produce forage for a cattle enterprise – mostly dairying – and a small sheep flock. With the shift to CAP conditions it will be more advantageous to sell barley as grain rather than process it through pigs. As a consequence, the model predicts a contraction in the size of the pig herd. With grazing livestock the main adjustments are a small expansion in the dairy herd and the disappearance of sheep. Some increase in the acreage of potatoes is expected, but otherwise the system of cropping hardly changes.

105

Cropping farms

Programming results for one cropping farm are given in Table 8.8. The emphasis in the cropping is on cereals, with intensive grain production the key to the farming system.

Table 8.8

Optimal Farm Organisation on One Cropping Farm
1969/70 and 1978/9

Type of farm	Cropping, mainly cereals	
Size of farm	328 acres	
Cropping (acres)	1969/70	1978/9
Winter wheat	130	133
Barley	130	133
Sugar beet	4	4
Maincrop potatoes	7	2
Stocking (numbers)		
Dairy cows	1	–
Beef cows	–	2
18-month beef	19	19
Summer-finishing beef stores	3	2
Barley beef	20	–
Ewes	28	22
Sows	4	21

Given the intensity of cereal production in the base-year programme there is little scope for expanding the acreage of wheat and barley grown on this type of farm, notwithstanding the substantial increase forecast for cereal gross margins. A marginal expansion in grains acreage is predicted, however, with a concomitant contraction in the area under potatoes.

A small residual acreage of grassland is utilised by beef cattle and sheep, the latter also drawing on stubbles and other arable by-products. Some substitution of land-using beef cattle for sheep is predicted by 1978/9. But the most interesting feature of the model so far as livestock is concerned is the disappearance of barley beef by 1978/9 – repeating the pattern on other farms – and a substantial increase in the size of the pig-breeding herd. This implies that with a change in the relative profitability of pigs and barley beef, cereal farmers will find it more profitable under EEC conditions to process their grain through pigs.

Farm incomes

The figures in Table 8.9 show the change in farm incomes predicted for eight farm models over the period 1969/70 to 1978/9.

Table 8.9

Change in Farm Incomes for Eight Farm Models
1969/70 to 1978/9

Farm type	Size	Change in money incomes	Change in real incomes
		Per cent	Per cent
Predominantly dairy	Small	+62	+14
Mainly dairy	Large	+77	+29
Cattle	Medium/large	+85	+37
Cattle and sheep	Medium/large	+53	+ 5
Sheep	Small/medium	+55	+ 7
Cereals	Medium	+58	+10
Mixed	Small	+45	− 3
Pigs and poultry	Medium/large	+16	−32

The results indicate that most farms – and particularly dairy and cattle farms – can expect substantial increases in incomes expressed in money terms over the ten-year period, but much of the improvement will be eroded by inflation. Real incomes of mainly dairy and cattle farms are predicted to increase over the period 1969/70 to 1978/9 by about one-third; real incomes on cereal, predominantly dairy, sheep and cattle and sheep farms may increase modestly, while on mixed farms and pig and poultry farms – especially the latter – the effects of inflation may more than offset any increase in money incomes, so that real incomes decline.

Farm structure

Structural change has been a feature of British agriculture over the last two decades. Farms have become fewer in number, larger in size and more specialised in their production patterns. As can be seen from Table 8.10, there has been a substantial decline in the number of full-time farms in England and Wales in recent years from 160,000 in 1963 to 145,000 in 1968, comprised for the most part of a fall in the number of small and, to a lesser extent, medium-sized farms, only partially offset by an increase in the number of large farms. Revision of the Standard Man Day values in 1968 introduced a major discontinuity into the series, but it is clear that the trend towards fewer and larger farms is continuing.

Table 8.10

Distribution of Holdings (a) in England and Wales by Size of Business Group, 1963, 1965, 1968 and 1970

('000)

Size group (standard man days)	1963	1965	1968[b]	1968[c]	1970
275–599	74.0	68.0	58.2	62.3	56.1
600–1,199	54.7	54.2	51.4	46.7	44.7
1,200–1,799				14.1	14.2
1,800–2,399	31.6	34.0	35.3	5.8	5.8
2,400–4,199				5.6	5.8
4,200 and over				3.0	3.2
Total full-time	160.3	156.2	144.9	137.4	129.7

Source: *The Changing Structure of Agriculture* (HMSO, London, 1970); and Agricultural Statistics.
[a] Holdings with 275 standard man days or more.
[b] Distribution based on 1967 SMD values for comparison with previous years.
[c] Distribution based on 1968 revised SMD values.

Given this fact, it was necessary to devise some method of projecting structural weights forward to 1978/9 to provide a basis for raising results for individual farm models to the aggregate level. The method used involved the estimation of a transition probability matrix P, which shows the probability that a farm, or unit of land, moves from state 'i' in one period to state 'j' in the next period. Having calculated such a matrix, the projection of numbers in succeeding periods is a matter of powering P and multiplying it by the base-year numbers in each state. The results of the projections are summarised in Table 8.11.

The number of full-time farms in England and Wales is projected to fall by over 24,000 between 1970 and 1980, thus continuing the trend described earlier for the 1960s. It is predicted that by 1980 there will be only 105,000 farms providing a full-time livelihood for their occupiers, compared with

Table 8.11

Projected Number of Farms in England and Wales by
SMD Size Groups, 1972, 1976 and 1980
('000)

Size group (SMD)	1972	1976	1980
275–599	51.9	44.9	39.2
600–1,199	42.2	37.8	33.6
1,200–1,799	14.1	14.3	14.3
1,800–2,399	6.0	6.3	6.5
2,400–4,199	5.9	6.5	7.2
4,200 and over	3.4	4.0	4.7
Total full-time	123.4	113.9	105.4

just under 130,000 in 1970. Within this overall decline changes in the distribution of farms between the various size categories will be occurring. The number of one- and two-man farms is projected to fall by 28,000 over the decade, but this will be offset by an increase in the number of large farms in the 1800 SMD and over-size groups.

Conclusions

Any conclusions that are drawn from the results of the eight farm models described above must be tentative, particularly since the models are not at this stage constrained by any aggregate restrictions. Nevertheless, if the results of these early models are representative of the type-group to which they belong, the trends that are shown are broadly consistent with the conclusions of earlier work undertaken in the Unit. Perhaps the most interesting of these is that there is no evidence of any marked expansion of the acreage of cereals grown in the UK following the adoption of EEC price levels. Despite the relative improvement in cereal gross margins, on only two farms was an increase in cereal acreage predicted. On the others there was no change in the total acreage of wheat and barley over the ten-year period. This confirms the conclusion reached by the author and Dr P. W. H. Weightman in a study completed some two years ago.[5] The reason for the relative stability of the cereal acreage is two-fold. First, there is the question of rotational constraints, which limit the area of cereals that can be grown on any one farm in the interests of good husbandry; this is particularly important in the major arable areas where the upper limit on cereal acreages has already been reached and, in some cases perhaps, exceeded. Second, and of relevance to livestock production areas, many farms in the UK have a comparative advantage in grazing livestock, the profitability of which will also improve under the CAP; for these farms it seems that few resources will be diverted to grain production.

On the supply side, some increase in grain production can be expected as yields improve, while on the demand side the results of a study recently completed by the Unit point to a decline in the consumption of grains commonly grown in the UK of between 1.7 and 2.9 million tons over the period 1969/70 to 1977/78.[6] This estimate was based on a decline in the usage of barley, oats and soft wheat for animal feed of 3.2 to 3.7 million tons, only partly offset by increases in usage for human food of 0.8 to 1.5 million tons. Radical policy changes apart, a surplus of UK-type grains may develop by the end of the transition period.

Turning to the other major farm crops – potatoes and sugar beet – little useful can be said until programming results are available for general cropping farm models. On the models completed to date, little change is predicted in the acreage of potatoes and sugar beet grown, but these farms account for only a very small proportion of the national total. So far as sugar beet is concerned, much depends on whether the UK government obtains an increased national quota when the Commonwealth Sugar Agreement comes up for renegotiation in 1974.

On the livestock side, some expansion in the national cattle herd is indicated. While the size of the national dairy herd may increase marginally, with a consequential small expansion in milk production, the emphasis will be on expanding the beef-breeding herd. This reflects the change in the relative profitability of beef and milk production in favour of beef. With a stable dairy herd and an expanding beef herd, a considerable boost to domestic beef production is probable; the increase could be of the order of 20 per cent. The structure of beef production will change, with less emphasis being put on intensive systems of the barley beef type because of increases in feed grain prices. At the moment it looks as if a small contraction will occur in the national ewe flock to accommodate the increased cattle population. An interesting aspect of the adjustments within the ruminant livestock sector is that, with little change in the acreage of arable crops, the increase in livestock numbers will be accomplished primarily through an intensification in the level of grassland management rather than through an increase in the forage area.

The prospects for farm incomes vary according to the type of farm. In money terms, farms where the emphasis is on cattle – both dairy and beef – seem likely to be favoured. Cereal farms should also receive a considerable boost in income. The prospects for sheep farms and mixed farms are slightly less promising, although still fairly good. Only on pig and poultry farms is a small income improvement predicted. In all cases, however, much of the benefit of higher money incomes will be offset by inflation.

References

1) Barnard, C. S., Casey, H. and Davey, B. H., *Farming Systems and the Common Market*, Agricultural Adjustment Unit Bulletin 5 (1968).
2) Davey, B. H. and Weightman, P. W. H., 'A Micro-economic Approach to the Analysis of Supply Response in British Agriculture', *Journal of Agricultural Economics*, XXII, 3 (1971), pp. 297–319.
3) Sturgess, I. M. and Reeves, R., *The Potential Market for British Cereals*, (Home-Grown Cereals Authority, May 1972), Table 5.8, pp. 5–9.
4) Separate models will be developed for Scotland and Northern Ireland at a later date.
5) Davey and Weightman, *op. cit.*
6) Sturgess and Reeves, *op. cit.*

9 Internal Agricultural Trade in an Enlarged EEC

Adrien RIES*
Head of Division
Commission of the European Communities, Brussels

Introduction

A widespread opinion on both sides of the Channel is that the basic objective of the CAP has been to ensure a fair standard of living for the agricultural population (Treaty of Rome, Article 39). This is certainly a correct assumption. However, it must be recalled that the major goal of the Treaty is to establish a Common Market between the six member countries. This Common Market has to be extended to agriculture and to trade in agricultural products. In other words, the basic objective of the CAP is the interregional division of labour.

In the long term, agricultural production should be located according to the principles of the law of comparative costs. Indeed there is no other economic activity where the local conditions of production (climate, soil and so on) are of such importance as in agriculture. Hence the interregional division of labour inside the Community means a reallocation of resources, a new production pattern and more specialisation on a Community level. As a result, the opening of the traditionally closed home markets of the Six must make room for a substantial increase in intra-Community trade.

The purpose of this paper is (a) to explain how the CAP pursues the basic goal of interregional division of labour; (b) to show the development of internal agricultural trade over the past decade; and (c) to speculate about some likely directions of trade within the enlarged Community.

The framework of the CAP and internal agricultural trade

To reach the basic goal of international division of labour and the general objectives of the CAP, the Treaty of Rome included as the major policy instrument the common organisation of agricultural markets. There are also some secondary instruments such as FEOGA, price controls and improvement of internal competition. By establishing their Common Market the member states intend to contribute to the harmonious development of world trade, the progressive abolition of restrictions on international exchange and the lowering of customs barriers (Article 110).

The Community has established market regulations for nearly all agricultural commodities (Annex II of the Treaty). Under the rules of the Common Market organisation, free movement of all these commodities is possible within the Community. The arrangements include:

(a) A system of common prices, expressed in units of account, which regulates the price level of agricultural commodities inside the Community and provides a proper protection against the world market by levies or tariff duties, (the Community preference);

(b) on the internal market, trade restrictions, such as duties or quantitative restrictions are strictly forbidden, as is the use of indirect taxation to manipulate trade (Articles 95–8);

(c) the free movement of production factors (labour and capital) has been established, and the freedom for national governments to intervene by protecting their own farmers or their own production limited by the common competition (Articles 92–5) or transport policy rules;

(d) artificial obstructions to the free movement of goods, in the form of health or technical legislation, have been banned, or will be in the near future.

This framework, including common prices in one single common market, has existed since 1967. But the instrument has not proved fully sound. There are still some deficiencies. In particular, there is a gap between the instrument's theoretical abilities to reach the goal of interregional division of labour and its practical workability.

So far as the field of this paper is concerned, at least three problem areas can be identified. First, the level of common prices, price relativities and prices guaranteed for major commodities freezes, in a certain sense, the existing production pattern and may lead to surpluses. Second, international monetary problems have inhibited or prevented the correct working of the common price system. Third, national governments have never stopped trying to neutralise by national measures the full effect of the CAP.

Two other factors have also influenced the development of the Common Market. The size of the Community in relation to the dimensions of the six home markets and the long distances between production and consumption areas reinforces the importance of transportation costs. Historical evolution, poor marketing, differences in consumption habits and obsolete production structures have increased the so-called marketing gap in the field of agriculture. Both these external factors have contributed to the delay in the economical accomplishment of market integration between the Six.

Trade developments in recent years

On a worldwide basis trade in agricultural commodities has lost some of its relative importance. In 1969, at $45,000 million it represented only 20 per cent of world trade as against 30 per cent in 1960. The Community accounted, in 1969, for 25 per cent of world agricultural imports ($11,500 million) and 7 per cent of its exports ($3,200 million). Since 1960, it has increased its imports by $3,200 million and its exports by $1,000 million. The EEC is the world's largest importer and, after the United States, the second largest exporter of agricultural commodities. The proportion of agricultural imports in the total imports of the Community has decreased regularly since 1960. In 1970 agricultural imports represented 29.2 per cent of total community imports, compared with 37.7 per cent in 1960. On the other hand, although exports of agricultural products increased by 65 per cent between 1958 and 1969, their importance in the total exports of the Community has decreased to less than 9 per cent in 1969. The EEC is definitely an industrial and not an agricultural economy.

Between 1958 and 1970 the internal agricultural trade of the Six has multiplied more than five-fold. This growth is not quite as large as that of trade in all commodities, which increased more than six-fold. In 1970 internal agricultural trade represented 15 per cent of total internal trade (Appendix I), but agriculture accounts for only 6.6 per cent of the Community GNP. Since 1958 the internal agricultural trade of the EEC has grown three times more quickly than trade with third countries. In 1970 internal trade was more than 50 per cent of total imports of food and agricultural commodities from the rest of the world; it surpassed exports to non-EEC countries by nearly 90 per cent. Between 1963 and 1970 the value of the total agricultural production of the Six increased by almost 5 per cent per annum. Internal agricultural trade increased by 14.8 per cent per annum during the period 1958–70, compared with 4.6 per cent for imports from, and 5.5 per cent for exports to, third countries. By 1970 the Community had reached a degree of self-sufficiency of nearly 90 per cent (Appendix II).

The growth of internal agricultural trade has been widely different for the several member states. Italy's imports have increased most rapidly, followed by France, the Netherlands, Belgium/Luxembourg (BLEU) and Germany. On the export side, French exports have increased most, followed by Germany, BLEU, the Netherlands and Italy. However, these indicators of growth in trade do not reflect the relative importance of the different countries in the internal trade. Germany has always been, and still is, the biggest importer (41 per cent in 1970 as against 51 per cent in 1958), followed by Italy (18 per cent in 1970). On the other hand, the Netherlands have been, and still are, the biggest exporters (34 per cent in 1970 as against 42 per cent in 1958); they are followed by France (28 per cent in 1970) and BLEU.

The development of internal trade also differs widely from commodity to commodity. The growth in trade for the major commodities over the period 1958–70 is shown in Table 9.1.

Table 9.1

Growth in Internal EEC Trade for Major Agricultural Commodities 1958–70

(1958=100)

Beef	1,790	Poultry meat	541
Coarse grains	1,684	Wine	437
Pig meat	1,524	Rice	350
Dairy products	896	Fruit and vegetables	310
Wheat	746	Oil cakes	309
Oil seeds	601	Eggs	96

There has been a spectacular development in trade in feed grains and pig meat on one side and in beef on the other. The first movement may be interpreted as the result of a complementary integration and a higher degree of specialisation of production, while the second shows the possibilities of market integration for commodities for which the larger market is not self-sufficient.

Trade between the EEC and the four applicant countries – the UK, Denmark, Norway[1] and Ireland – increased between 1963 and 1970 more for all commodities than for agricultural products. In 1963, agricultural products still represented nearly 20 per cent of the total imports of the Six from the Four; by 1970 this had been reduced to 12 per cent. On the export side, the relative importance of agricultural products in exports from the Six to the Four decreased from 16.8 to 11.5 per cent between 1963 and 1970 (Appendix III). Over the same period the weight of the Four in total Community trade with third countries has decreased not only for agricultural products, but also for all commodities. The simultaneous development of the EEC and of EFTA seems to have led towards different patterns of market integration. Of the four applicant countries, only the UK imports large quantities of agricultural products from the Community; in 1968 nearly 11 per cent of its food and agricultural commodities came from the Six. Denmark, followed by the UK, are the main suppliers of the EEC.

Trade between the four applicant countries increased by 67 per cent for all commodities, but only by 19 per cent for agricultural products, during the period 1963–9. Imports of agricultural products represented 28 per cent of total imports in 1969 compared with 37 per cent in 1963. The UK is the chief importer of agricultural products from the Four (85 per cent in 1969), mainly from Denmark and Ireland. On a worldwide basis the four applicant countries imported, in 1969, 7,500 million units of account of agricultural products. This is roughly the same amount as in 1963. In relation to this total trade, internal agricultural trade between the Four represents only 16 per cent (1,037 million u.a.), but it still surpasses trade with the Six by 30 per cent (Appendix IV).

The development of internal agricultural trade shows that a degree of market integration has occurred between the Six during the last decade. It may be concluded that Community preference has played an important role since intra-EEC trade developed three times more quickly than extra-EEC trade. There has also been a movement towards a better division of labour inside the Community. Crop production expanded in France, livestock production in the consumption area of northern Europe (Appendix V). Full market integration, as a consequence of the optimum allocation of resources, is a matter of long-term structural evolution; its achievement also needs the full integration of the economies of the Six.

Likely directions of trade within the enlarged community

The enlargement of the Community will modify the agricultural economy of the Common Market. The agricultural area will increase by 6.8 million square kilometres, with the arable area expanding by 21 million hectares and pastures by two-thirds. The structure of holdings will improve with average farm size becoming larger. The consumption market increases by 70 million people to reach a total of 260 million consumers. A static analysis of the consequences of the enlargement of the Community suggests that the degree of self-sufficiency of the enlarged Community will decrease, especially for those commodities like wheat, sugar and butter where the Community of Six is in surplus (Appendix II). In addition, the complementarity of the agricultural economies appears quite clearly. The figures in Table 9.2 show the surpluses (+) and deficits (−) of the EEC and the four applicant countries for selected commodities in 1969.

On the other hand, there are two groups of products where no complementary effect of enlargement is to be expected. There is an addition to surpluses for pig meat, eggs and poultry and an addition to deficits for coarse grain, fats and oils.

A realistic analysis of future trade trends requires a dynamic approach.

Table 9.2

Surpluses and Deficits of Selected Agricultural Products in the EEC and the Four
Applicant Countries 1969

| | million tons | |
Commodity	The Six	The Four
Wheat	+6	−4.3
Sugar	+1.4	−1.9
Butter	+0.20	−0.23
Vegetables	+0.4	−0.9
Fruit	–	−1.7
Beef	−0.55	+0.35

Source: 'Les aspects agricoles de l'élargissement de la communauté écono-
mique européenne; l'influence de l'intégration et de la coopération écono-
miques internationales sur l'agriculture belge', *Chronique de politique
étrangère*, XXIII, 3-5, (May–September 1970).

Several changes could affect agriculture in the applicant countries. The
progressive application of the CAP (Community preference, higher prices
except in Norway, disappearance of production limitations) could liberate
the hidden forces of the most efficient agricultures in Europe. There is no
reason why production, especially of products with relatively high price
incentives such as wheat and cattle, should not expand. On the other hand,
changes in consumer prices may become substantial after enlargement.
Consumer prices of butter and cheese may go up substantially, as may those
of wheat and feedgrains. As a result, and in the long term, these price changes
may considerably alter the levels and patterns of demand and supply of
agricultural products in the new member countries. The opinions of those
who have studied the likely evolution of production and consumption of
agricultural products in the enlarged Community diverge widely. Especially
in the case of the UK, the question of a large expansion of grain production
or of cattle and milk production is of vital importance for the future development of internal trade. So is the assumption made on the development of
the CAP after enlargement on such important issues as price levels, price
relativities, Community preference and structural policy, including aids to
hill farming.

While it is very difficult to predict the long-term development to the
1980s and 1990s, it is possible to make a few observations about the evolu-

tion during the transitional period from 1973 to 1978. It may be assumed that, during this period, trade diversion will take place only exceptionally and in a limited scope. The provisions of the Brussels Treaty freeze for several years existing UK trade relations for sugar with the Caribbean and other developing Commonwealth countries, and for butter with New Zealand. The existing trade relations between the Six and the Four, the consumption habits of British housewives and the Channel crossing are just some of the numerous elements affecting trade in fresh vegetables and fruit within an enlarged Community. Finally, there is no reason why the traditional dependence of Danish agriculture on the British market should not continue in the future, even if there is some trade diversion from Denmark to Germany and from Benelux to the UK.

In 1970 internal agricultural trade (imports) between the ten countries of the enlarged Community reached 9,000 million units of account. The future development of internal trade may be assumed to lie between two extremes, namely a maximum increase of 14.5 per cent per annum, which has been the trend of internal agricultural trade in the Community of Six over the past 12 years and a minimum increase of 10 per cent per annum, which has been the trend of agricultural trade between the Ten over the past years. These assumptions lead to a projection of internal agricultural trade in the enlarged Community of 14,400 to 17,600 million units of account in 1975.

Conclusion

The goal of an interregional division of labour will continue to be the basis of the CAP in the enlarged Community. There is no doubt about the high level of complementarity between the agricultures of the Six and the Four. There is no doubt either of the high level of efficiency of the bulk of agriculture in the new member countries. But the way in which adaptation to enlargement will take place is not easy to predict, more particularly as there is uncertainty as to the future development of the CAP. In any case, the future of internal agricultural trade will depend on external factors, such as changes in consumption habits in the UK and in transportation facilities for crossing the Channel.

The establishment of an enlarged Community raises problems in international trade relations, for example trade diversion against third countries, which increases the need for international arrangements. On the other hand, the progressive realisation of interregional division of labour will create more problems in many regions of the Community. A sound regional economic development policy must, therefore, accompany the future development of the CAP.

Reference

1) This paper was finalised before the result of the Norwegian referendum was known.

Note

* The views expressed in this paper represent the personal opinions of the author. They in no way represent the official opinion of the European Communities.

Appendix IA
Internal Trade of the EEC: Imports – all Products

Year	Value ($ million)[a]					
	EEC	France	Bleu	Nether-lands	Germany	Italy
1958	6786.4	1227.4	1461.6	1517.9	1896.1	683.6
1959	8081.8	1361.9	1620.9	1749.5	2460.6	889.0
1960	10151.2	1847.6	1893.8	2076.1	3023.9	1309.8
1961	11713.5	2101.8	2135.4	2514.0	3427.2	1535.0
1962	13412.2	2522.7	2323.9	2683.1	3995.1	1887.4
1963	15708.5	3125.7	2684.4	3081.7	4341.9	2474.8
1964	18050.0	3762.2	3154.8	3671.1	5097.3	2364.6
1965	20425.5	4015.2	3473.4	3985.1	6660.4	2291.4
1966	22918.4	4852.3	4008.6	4331.8	6938.5	2787.2
1967	24160.7	5373.7	3983.6	4546.1	6867.5	3389.9
1968	28383.8	6616.6	4553.3	5146.2	8358.5	3709.2
1969	36334.5	8690.2	5733.9	6130.0	10862.1	4818.3
1970	42801.8	9255.8	6685.4	7483.2	13231.9	6146.2

Year	Index (1958=100)					
	EEC	France	Bleu	Nether-lands	Germany	Italy
1959	119.1	111.0	110.9	115.3	129.8	130.0
1960	149.6	150.5	129.6	136.8	159.5	191.6
1961	172.6	171.3	146.1	165.6	180.7	224.5
1962	197.6	205.5	159.0	176.8	210.7	276.1
1963	231.5	254.7	183.7	203.0	229.0	362.0
1964	266.0	306.5	215.2	241.8	268.8	345.9
1965	301.0	327.2	237.6	262.5	351.3	335.2
1966	337.7	395.4	274.3	285.3	365.9	407.7
1967	356.0	437.8	272.6	299.5	362.2	495.9
1968	418.2	539.1	311.5	339.0	440.8	542.5
1969	535.4	708.0	392.3	410.4	572.9	704.8
1970	630.7	754.1	457.4	493.0	697.8	899.1

Source: *Monthly Statistics of Foreign Commerce*, Statistical Office of the European Community.
[a] New exchange rate: March 1961 for the Netherlands and Germany, August 1969 for France, October 1969 for Germany.

Appendix IB
Internal Trade of the EEC: Imports – Food and Agricultural Products

Year	Value ($ million)					
	EEC	France	Bleu	Nether- lands	Germany	Italy
1958	1246.1	126.9	226.2	128.3	636.1	128.6
1959	1546.2	180.7	262.9	149.9	786.4	166.2
1960	1785.3	211.6	·280.0	167.8	898.5	227.2
1961	1967.3	202.4	303.7	213.3	1028.1	219.9
1962	2220.9	264.1	321.1	205.8	1183.9	245.7
1963	2489.8	351.4	367.7	227.2	1178.4	365.1
1964	2821.6	432.5	419.8	268.3	1300.4	400.6
1965	3335.5	468.4	489.2	316.9	1614.1	446.9
1966	3599.7	485.6	532.3	347.0	1713.5	521.3
1967	3848.9	553.7	561.8	401.0	1696.5	636.0
1968	4557.2	709.8	675.7	497.7	1901.4	772.7
1969	5800.3	923.4	823.4	701.7	2390.2	961.6
1970	6516.4	978.7	962.8	741.2	2678.1	1155.6

Year	Index (1958=100)					
	EEC	France	Bleu	Nether- lands	Germany	Italy
1959	124.1	142.4	116.2	116.8	123.6	129.2
1960	143.3	166.7	123.8	130.8	141.3	176.7
1961	157.9	159.5	134.3	166.3	161.6	171.0
1962	178.2	208.1	142.0	160.4	186.1	191.1
1963	199.8	276.9	162.6	177.1	185.3	283.9
1964	226.4	340.8	185.6	209.1	204.4	311.5
1965	267.7	369.1	216.3	247.0	253.7	347.5
1966	288.9	382.7	235.3	270.5	269.4	405.4
1967	308.9	436.3	248.4	312.5	266.7	494.6
1968	365.7	559.3	298.7	387.9	298.9	600.9
1969	465.5	727.7	364.0	546.9	375.8	747.7
1970	522.9	771.2	425.6	577.7	421.0	898.6

Source: *op. cit.*

Appendix IC
Internal Trade of the EEC: Exports – all Products

Year	Value ($ million)					
	EEC	France	Bleu	Nether-lands	Germany	Italy
1958	6864.0	1135.6	1377.3	1336.8	2406.0	608.3
1959	8167.5	1523.6	1524.1	1597.2	2730.1	792.5
1960	10245.7	2041.6	1907.8	1849.3	3369.1	1077.9
1961	11898.9	2419.7	2089.4	2050.6	4025.9	1313.3
1962	13563.7	2711.5	2458.4	2256.1	4512.5	1625.2
1963	15926.1	3091.5	2942.3	2647.2	5451.9	1793.2
1964	18382.9	3487.2	3498.6	3233.1	5909.7	2254.2
1965	20822.2	4114.7	3947.1	3561.3	6306.3	2892.8
1966	23233.9	4608.4	4295.8	3749.9	7318.0	3261.7
1967	24512.8	4701.5	4432.9	4002.9	8002.5	3373.0
1968	28910.3	5452.1	5249.1	4790.2	9339.8	4079.1
1969	36464.5	7118.2	6799.9	5991.9	11571.0	4983.5
1970	43302.7	8661.5	7952.3	7289.7	13726.7	5672.5

Year	Index (1958=100)					
	EEC	France	Bleu	Nether-lands	Germany	Italy
1959	119.0	134.2	110.6	119.5	113.5	130.3
1960	149.3	179.8	138.5	138.3	140.0	177.2
1961	173.3	213.1	151.7	153.3	167.3	215.9
1962	197.6	238.8	178.5	168.8	187.6	267.2
1963	232.0	272.2	213.6	198.0	226.6	294.8
1964	267.8	307.1	254.0	241.9	245.6	370.6
1965	303.4	362.3	286.6	266.4	262.1	475.6
1966	338.5	405.8	311.9	280.5	304.5	536.2
1967	357.1	414.0	321.9	299.4	332.6	554.5
1968	421.2	480.1	381.1	358.3	388.2	670.6
1969	531.2	626.8	493.7	448.2	480.9	819.3
1970	630.9	762.7	577.4	545.3	570.5	932.5

Source: *op. cit.*

Appendix ID
Internal Trade of the EEC: Exports – Food and Agricultural Products

Year	Value ($ million)					
	EEC	France	Bleu	Nether- lands	Germany	Italy
1958	1212.9	221.7	167.2	507.6	101.8	214.5
1959	1525.8	318.3	201.6	631.4	120.0	254.7
1960	1775.1	453.7	219.1	679.5	135.1	287.6
1961	1965.3	529.8	252.8	716.0	137.3	329.4
1962	2199.7	551.6	311.9	783.4	153.6	399.2
1963	2480.2	680.5	372.4	877.9	182.9	366.5
1964	2778.7	783.2	385.6	988.3	217.3	404.3
1965	3337.5	924.0	472.4	1163.7	262.0	515.5
1966	3509.6	1028.0	490.4	1180.0	292.8	518.4
1967	3882.8	1105.5	563.0	1291.1	417.7	505.5
1968	4616.4	1356.4	668.6	1552.1	533.5	505.8
1969	5825.5	1891.0	828.9	1844.6	652.2	608.8
1970	6531.1	1941.5	941.5	2239.3	758.9	649.8

Year	Index (1958=100)					
	EEC	France	Bleu	Nether- lands	Germany	Italy
1959	125.8	143.6	120.6	124.4	117.9	118.7
1960	146.3	304.6	131.0	133.9	132.7	134.1
1961	162.0	239.0	151.2	141.0	134.9	153.6
1962	181.3	248.8	186.5	154.3	150.9	186.1
1963	204.4	306.9	222.7	172.9	179.7	170.9
1964	229.1	353.3	230.6	194.7	213.5	188.5
1965	275.2	416.8	282.5	229.3	257.4	240.3
1966	289.4	463.7	293.3	232.5	287.6	241.7
1967	320.1	498.6	336.7	254.4	410.3	235.7
1968	380.6	611.8	399.9	305.8	524.1	235.8
1969	480.3	853.0	495.8	363.4	640.7	283.8
1970	538.5	875.7	563.1	411.2	745.5	302.9

Source: *op. cit.*

124

Appendix II

Degree of self-sufficiency of the community in agricultural products, 1970–1

Products	The Six	The Ten
A *All Cereals*[1]	86	80
of which		
Soft wheat	102	84
Hard wheat	74	84
Rye	94	92
Barley	91	91
Oats	88	91
Maize	66	56
Other cereals	16	47
B *Rice*	134	–
C *Sugar*[2]	106	84
D *Wine*[3]	104	–
E *Fats and Oils* (total)[4]	40	–
of which		
Vegetable fats and oils	23	–
Slaughtering oils	84	–
Oils and fats from sea animals	5	–
F *Meat*[5]		
Beef	88	94
Pig meat	101	106
Poultry meat	101	103
G *Dairy Products*[6]		
Liquid whole milk	100	100
Liquid skimmed milk	100	100
Butter	104	86
Cheese	102	100
Powdered milk	133	129
Condensed milk	154	148
H *Fruit and Vegetables*[7]		
Fresh vegetables (including canned)	99	–
Fresh fruit (including canned fruit and juices)	88	–

Citrus fruits (including canned fruit and juices)	52	–

Source: Agricultural Statistics, Statistical Office of the European Community.

[1] Marketable production as per cent of total domestic consumption.
[2] Net production as per cent of total domestic consumption.
[3] Production as per cent of total domestic consumption.
[4] Domestic production as per cent of available supplies.
[5] Domestic net production as per cent of available supplies, without fat.
[6] Production as per cent of available supplies.
[7] Marketable production as per cent of total domestic consumption.

Appendix III
Trade between the community and the four applicant countries

	1963 $ million	1970 $ million	Index 1970 1963=100
I EEC Imports			
All commodities	3312	5639	170
Food and agricultural products	649	702	108
of which come from			
United Kingdom	216	258	119
Denmark	350	343	98
Norway	57	89	156
Ireland	26	32	123
II EEC Exports			
All commodities	3338	6150	184
Food and agricultural products	562	706	125
of which go to			
United Kingdom	456	552	121
Denmark	60	85	142
Norway	32	47	147
Ireland	14	22	157

Source: *Tableaux Analytiques CST*, Statistical Office of The European Community.

Appendix IV
Internal trade of the four applicant countries

	1963 $ million	1969 $ million	Index 1969 1963=100
I Imports			
All commodities	2329	3756	167
Food and agricultural products	873	1037	119
of which come from			
United Kingdom	98	118	120
Denmark	433	467	108
Norway	47	85	181
Ireland	295	367	124
II Exports			
All commodities	2193	3471	158
Food and agricultural products	843	989	117
of which go to			
United Kingdom	708	831	117
Denmark	19	30	158
Norway	43	45	105
Ireland	73	83	114

Source: *Statistical Bulletins, Foreign Trade*, OECD.

Appendix V
Indices of the volume of agricultural production in EEC

1963=100

Production	1968	1969	1970
A *Final total production*			
Germany	117.5	118.1	118.7
France	121.8	117.7	124.4
Italy	120.8	125.2	125.8
Netherlands	128.4	133.6	142.3
Belgium	122.5	126.5	134.3
Luxembourg	100.1	98.8	102.4
EEC	120.8	121.3	124.9
B *Final crop production*			
Germany	111.6	106.7	109.6
France	129.5	122.1	136.1
Italy	112.2	117.2	116.4
Netherlands	133.2	134.1	140.5
Belgium	121.5	123.2	122.6
Luxembourg	81.9	80.5	94.6
EEC	118.9	117.8	122.7
C *Final animal production*			
Germany	117.5	120.8	123.5
France	117.2	115.0	117.3
Italy	136.8	140.1	143.4
Netherlands	125.7	133.3	143.3
Belgium	123.0	128.2	140.6
Luxembourg	107.6	106.2	105.6
EEC	121.4	123.0	126.7

Source: *Yearbook of Agricultural Statistics 1971*, Statistical Office of the European Communities.

10 Social Problems of Rural Communities

D. B. WALLACE and P. J. DRUDY
Department of Land Economy,
University of Cambridge

Introduction

Consideration of the effect of entry into the EEC upon agriculture is not the principal aim of this paper. Rather it is a 'background' paper, which sets out to describe conditions in rural communities that have existed for some time and to consider the probable causes and some of the effects thereof. Once these are understood, it is possible to hypothesise what the effect of joining the EEC may be. In the main, however, that is the prerogative of Professor Gaskin's paper. Here it is proposed to concentrate rather on the analysis of existing problems.

For present purposes, the term 'rural communities' can be loosely defined as any geographical area that is mainly under some form of agricultural, or forestry, use and contains no large towns. The human population may be grouped into scattered settlements, some large enough to be counted as small market towns, but in some cases there may be few real centres of population, and many people may live in small family groups dispersed through the countryside. Nevertheless, even in the areas of most dispersed population there are usually one or more recognised centres both of local trade and, more important, social administration.

It should be made clear that the paper is not concerned with isolated settlements that have, or have had, a mainly industrial origin, such as Tow Law in County Durham or Blaenau Festiniog in Wales, nor with rural settlements that exist beyond the green belts of large urban areas, and are now mainly inhabited by commuters who travel away from the immediate area each day to work. It is difficult to frame a watertight definition of the type of area under examination that will satisfy all social scientists. Nevertheless it is hoped that, by offering examples throughout the analysis that follows, a common sense impression will be given of the type of settlements being examined.

Two major areas are located respectively in mountain areas of Scotland and Wales. In the former, the area is that covered by the Highlands and Islands Development Board and includes the counties of Argyll, Caithness, Inverness, Orkney, Ross and Cromarty, Sutherland and Shetland Islands

(Zetland), with a population of about 280,000 people. The only sizeable town is Inverness, with a population of around 30,000, the next in size being less than 10,000. The area is predominantly one of mountain farming, relying on sheep in the main, with subsidiary cattle where possible. Small holdings or crofts predominate, 80 per cent of all holdings having less than 30 acres of crops and grass, excluding rough grazing. Mid-Wales is also an area devoted mainly to sheep farming, though cattle are also significant. Roughly one-third of the farms are less than 30 acres, a far smaller proportion than in the Highlands. The area comprises the counties of Cardigan, part of Brecon, Merioneth, Montgomery and Radnor, with a population of about 175,000. The largest town is Aberystwyth with a population of about 10,000, with a few small market towns at the 5,000 level.

General data for these areas are analysed below, and for two smaller areas, North Mayo and North Norfolk. For the latter area some new information is available from field work carried out by the authors. These two areas contrast with the first two, not only in size, but as areas that are farmed more intensively. North Norfolk contains three small market towns with populations under 5,000 each. Otherwise the population, amounting in total to some 30,000 lives either in small villages, often of under 300 people, or scattered through the countryside on farmsteads. The farming is almost entirely arable and the major crops are barley, wheat and sugar beet, in that order. While a surprising number of farms, about 40 per cent of the total, are under 30 acres, 15 per cent are over 500 acres and account for a majority of the acreage. North Mayo is an area of mainly grass farms, many of them on the hills. There is only one market town of about 6,000 people, with the bulk of the population of 30,000 living outside the main settlements, either in small villages or on the farms themselves. The great majority of the farms are pastoral, with very little emphasis on either cereals or roots. In recent years, with a reduction in the number of working horses, the need for oats, the main cereal, has declined, and the grass acreage has increased, together with the number of cattle, especially dairy cows. Between 1962 and 1970 the production of milk in the whole county increased from two and a half to over seven million gallons, and the number of suppliers increased from just under 2,000 to 3,300. The holdings themselves are small, 70 per cent being under 30 acres and only 4 per cent over 100 acres. It is hoped that the concept of 'rural community' as used in this paper will be understood from these examples.

The basic social problem

One problem that is common to the areas under study is that of declining total population. Table 10.1 shows population figures for each intercensal period from 1951 to 1971 in these areas.

During the period under review all four areas have recorded a decline in population, although the Scottish area showed some recovery during the last ten years. The statistics for both Scotland and Mid-Wales conceal a number of wide inter-county variations. The decline of 1 per cent for the Scottish area as a whole is accounted for solely by an increase of 22.2 per cent in Caithness and 5.4 per cent in Inverness. The other five counties registered declines ranging from 3.7 per cent for Ross and Cromarty to 19.7 per cent for Orkney. In the Mid-Wales area, the only county to record an increase was Cardiganshire (2.9 per cent) and this was due almost entirely to the growth of Aberystwyth and its hinterland. All the other counties registered declines ranging from 7 per cent for Montgomery to 14 per cent for Merioneth. Such inter-county variation is one argument for examining smaller statistical units such as North Norfolk or North Mayo, where the true trends are clearly discernible.

Agricultural adjustment and labour mobility

The areas selected have one predominant common characteristic, namely the importance of agriculture as a major occupation for the adult population, especially the males. Therefore, any factors that have caused changes in either the structure of agriculture or the pattern of labour use must have a much more marked effect on the population of the areas than would be the case where agriculture was not such an important occupation.

During the last twenty years agriculture throughout the UK has been under strong economic pressures, with production and productivity increasing more rapidly than prices. During this period real prices per unit for most agricultural products have fallen and, in their attempts to sustain their

Table 10.1

Population Change in Selected Communities 1951–71

Area	1951	1961	1971	Actual change 1951–71	Percentage change 1951–71
Highlands and Islands of Scotland	285,786	277,948	282,901	− 2,885	− 1.0
Mid-Wales	185,729	178,546	174,089	−11,640	− 6.3
North Norfolk	35,518	32,141	29,385	− 6,133	--17.3
North Mayo	40,030	35,126	31,638	− 8,392	−20.9

incomes, farmers have had to concentrate their efforts on raising productivity in relation to their fixed inputs, and especially in relation to themselves and also their permanent labour force. Indeed, rising productivity was expected of the industry and in the Annual Price Reviews a substantial element of increased costs of the order of £25–30 million per annum was expected to be absorbed by increased efficiency on the part of the industry, before any recoupment of increased cost was contemplated.

One effect of increasing efficiency can be seen from Table 10.2.

Table 10.2

Agricultural Labour, 1958 and 1968*

Area	1958 Total regular workers	1968 Total regular workers	Change	
			Actual	Per cent
North Norfolk	3,285	1,980	−1305	−40
Highlands and Islands	9,189	4,826	−4363	−47
Mid-Wales	8,494	4,924	−3570	−42

* Excluding farmers.
Source: *Agricultural Statistics*, MAFF.

Taking the case of North Norfolk first, it might be expected that in an arable area, and one with a fairly considerable hired labour force, the impact of such pressures would be marked. The introduction of larger machines, especially for cultivation and the harvesting of labour-intensive crops such as sugar beet and potatoes had a marked effect on the demand for labour, and the hired labour force fell by 1,300 workers, a reduction of 40 per cent. What is perhaps more unexpected is the commensurate reduction in the permanent hired labour force in both the mountainous areas. Admittedly the actual numbers involved are smaller, bearing in mind the much larger geographical area, just over 4,000 in the Highlands and just under that figure in Mid-Wales, but the percentage changes were virtually the same at 47 and 42 per cent respectively. The movement out of agriculture has been brought about not only by technological change and economic pressures. A survey carried out on 153 farms in the North Norfolk area during 1970/1 illustrates that a combination of push and pull factors explains the movement best (Table 10.3).

132

Table 10.3

Reasons for Movement from North Norfolk Farms, 1960–70*

	Number	Per cent of total
Dissatisfied with work or conditions	86	18.0
Worker's wife or family dissatisfied	10	2.1
Attracted by other employment	24	5.0
Wages	111	23.3
Dismissed	35	7.3
Ill-health	18	3.8
Housing problems	18	3.8
Redundancy	154	32.3
Other reasons	21	4.4
Total	477	100.0

* Active workers only: workers who died or retired are not included.

The figures indicate that redundancy was the greatest single reason for leaving, accounting for 32.3 per cent of the movement. The basis of this redundancy has been discussed earlier and needs little further elaboration. Wages and dissatisfaction with farm work are the two other important factors in the out movement. It is not surprising that a high proportion moved for wage reasons. The relative position of agricultural workers is low compared to workers in other industries and there has been no gain relative to other industries in the past thirty years.[1]

No similar details are available for the other areas, but it is reasonable to assume that, apart from redundancy, similar 'pull' factors are operating. Indeed the increase in the population of Caithness is probably due principally to by the construction of the atomic power station at Dounreay and it is likely that such construction work was a significant draw to farm labour.

In the last ten years, in each area, there was a very sharp reduction in the paid employment available locally in a major local occupation. Not only would the net reduction have a severe effect, especially on those who lost their jobs, but the reduced employment available would have a depressing effect on school-leavers coming on to the labour market, especially as a farm job would often be the first employment taken up by boys, even if they moved on eventually to other occupations.

The economic pressures on agriculture did more than change the pattern of paid labour use. The reduction in real return per unit of produce meant that there was increasing pressure on another fixed resource, the farmer himself, to increase his turnover. There were many ways to do this, but one long-term solution was to operate a bigger unit. As the amount of land is the most fixed factor of all, this could come about only by a reduction in the number of units and expansion in their size, not only in absolute terms of acreage, but in turnover as well. The economic pressures were obviously severe on the smallest units, many of which were not intended to provide a full-time living off the land alone, but to be farmed alongside other employment, perhaps on larger farms, although in some areas, such as parts of Wales, with work in other industries, such as quarrying. These small units would prove difficult to maintain if the opportunities to earn off-farm incomes declined. In addition, on the small units the lower real unit price meant that it was now more difficult to earn a full income from a small acreage even where that acreage had been sufficient for a full-time income in the past. The effect of these pressures can be seen from Table 10.4.

Table 10.4

Number of Agricultural Holdings, 1958 and 1968

Area Size in acres	North Norfolk			Highlands and Islands*			Mid-Wales*		
	1958	1968	Change Per cent	1958	1968	Change Per cent	1958	1968	Change Per cent
1–29	436	264	− 39	22055	17495	− 21	7171	4184	− 42
30–99	204	144	− 29	3470	3206	− 8	6119	5293	− 14
100–299	155	138	− 17	1025	1102	+ 8	2808	3009	+ 7
300 and over	142	141	–	192	230	+20	129	216	+67
Total	937	687	− 27	26742	22033	− 18	16227	12702	− 22

* In these areas, the acreage size groups relate to crops and grass acres only.
Source: *Agricultural Statistics*, MAFF.

In all three areas the total number of holdings declined over the ten-year period from 1958 to 1968 by about a fifth. The bulk of the reduction has been in the smallest size group, as would be expected, but the proportionate reduction in the next size group has also been appreciable. In the two mountain areas the bulk of the increases in unit size lie in the remaining two groups, between 100 and 300 acres, and over 300 acres. In the case of Norfolk, uniformity of tabular presentation conceals the fact that it is only in the over-500-acre group that a net increase has taken place, so that in this arable area the obvious economies of scale that can accrue on a mechanised farm

134

have tended to lead to much bigger units than in the pastoral mountain areas, where rough grazings have been omitted from the figures.

The conclusion is that economic pressures have led in the past ten years to a sharp reduction in the number of small farms and a corresponding increase in bigger units. As farmers have maintained or increased total production, despite the reduction in the paid labour force, it must mean that there will be fewer job opportunities for the rural population in the future.

Community satisfaction and social provision

Apart from lack of job opportunities, there are other important variables that influence migration patterns. Level of satisfaction with the local community is one such variable. A survey of post-primary school-leavers carried out during 1971 illustrates the importance of this in the Norfolk area. Of 92 respondents who planned to migrate after completing school, 33.7 per cent had a low level of community satisfaction, 65.2 expressed a medium level of satisfaction and only 1.1 per cent had a high community satisfaction level. A high community satisfaction level is inversely related to the intention to migrate. One aspect that might be expected to influence the level of community satisfaction is the level of social provision of water supplies and sanitary facilities. In this respect, the rural regions compare unfavourably with the national picture. The position in two counties in Mid-Wales and the Highlands, together with North Norfolk, is presented in Table 10.5. The figures speak for themselves and need little elucidation.

It can be argued that there are a number of slum areas in our major cities where the dearth of such basic household facilites is a more serious problem

Table 10.5
Household Amenities in Selected Areas, 1966

| | Percent | | |
	No hot-water tap	No bath	No water closet
England and Wales	12.5	15.0	1.8
Montgomery	28.9	35.5	22.2
Radnor	24.6	28.8	18.0
Shetland Islands	37.5	38.8	22.2
Ross and Cromarty	19.8	22.9	14.2
North Norfolk	31.9	33.7	22.0

Source: Sample Census 1966.

than in some of the rural areas studied here. For example, in the London borough of Islington, 36.6 per cent of the households have no hot-water tap and 34.4 per cent have no bath; only 0.4 per cent lack a water closet. Similarly, 34.9 per cent of the households in the London borough of Tower Hamlets have no hot-water tap and 40.8 per cent lack a bath. Again, they are well provided with water closets, as only 0.3 per cent lack this facility. The absolute number of households involved in the London area is, of course, much greater and the problem more concentrated than in any of the rural areas referred to above. For example, 30,900 households lack a hot-water tap in Islington compared with 1,970 in the Shetland Islands. Similarly, 29,190 households are without a bath in Tower Hamlets compared with 4,770 in Montgomery. The magnitude of the problem in these urban areas is not in question, but to say that such city regions should receive priority at the expense of rural areas may also be a short-sighted approach. A slowing down in amenity provision in rural areas could conceivably accelerate the migration process to such urban areas and merely magnify the problem to un manageable proportions. In addition to this, it must be borne in mind that the London examples are exceptional and are being overcome under slum clearance programmes. In general, then, rural areas compare unfavourably with urban ones and with the overall national picture.

Provision of retail services

So far the discussion of social provision has centred on amenities and facilities provided in the main by local authorities. But one other aspect of rural isolation is the reputed lack of retail services. It is often said that a rural settlement has not only poorer provision in the range of shops, but a much lower density in relation to population. It is difficult to quantify this aspect in unequivocal terms and to avoid value judgements, but the general information contained in the 1961 Census of Distribution and Other Services throws some light on the problem. In Table 10.6 the position in the largest town in or near the areas examined is set out, together with information on one or more smaller market towns from each area, and then two or three rural districts.

For each settlement, the number of retail establishments per 1,000 population is shown, and then the average turnover of those establishments. Some of the trade of the establishments in both types of town is due to the influx of buyers from the rural areas, while little or none of the trade of the rural establishments will be due to the townspeople. Yet if a rural dweller has to travel to town for some of his requirements, then this can be regarded as one drawback to a rural settlement. Similarly, if the number of establishments in relation to the population, and also their actual size in terms of

136

Table 10.6

Retail Businesses in Selected County Towns and Rural Districts

Town/R.D.	Population	Number of establishments per 1,000 population	Average turnover per establishment £000's
Norfolk			
King's Lynn	27,554	15.7	17.9
Dereham	7,197	16.0	15.3
Docking	18,287	8.5	9.5
Walsingham	20,886	8.8	10.5
Highlands			
Inverness	29,773	10.9	24.3
Wick	7,397	15.4	13.7
Caithness North	1,527	8.7	6.9
Lochaber	10,016	4.8	13.5
Golspie	3,126	10.0	14.6
Mid-Wales			
Aberystwyth	10,418	14.8	19.8
Welshpool	6,332	17.3	10.3
Tregaron	4,805	11.0	4.0
New Radnor and Paincastle	3,786	3.4	5.6

Source: Census of Distribution and Other Services, 1961.

business done are less, these are also indications of the disadvantages of isolation. This is a very crude generalisation and some index of the range and size of retail and amenity provision, as suggested by Bracey[2], would be much better, but this is not available for these areas. From Table 10.6 it is clear that while the number of establishments per 1,000 population does not differ greatly between the main centre and the small market towns, a figure of about fifteen being observed in most cases, the turnover per establishment in the small market town is usually substantially less. When the rural districts are considered, not only is the number of establishments per 1,000 population usually much smaller the turnover is also much lower than for the market town, a notable exception being observed in the Scottish section, where two of the districts quoted provided evidence of turnover on a par with that of the market town.

The evidence above can be offered only as a very general guide, and the implications would be much easier to discern if the trend over time of changes in the ratio of establishments in rural areas, compared to that in towns, were known. But the evidence gives some support to the view that isolation of rural areas carries a definite disadvantage in the availability of retail services immediately in the area.

The effects of depopulation

Population decline has had a number of serious effects in the areas under study. Since migrants tend to consist mainly of young people of working age, a community of very young and ageing members remains behind. The age structures of the communities under study are compared to that of England and Wales in Table 10.7.

Table 10.7

Percentage Age Distribution of the Population in
England and Wales and Selected Communities, 1966

Age group	England and Wales	Highlands and Islands	Mid-Wales	North Norfolk	North Mayo
0–14	23.0	24.5	22.1	24.1	31.6
15–24	14.5	12.5	13.6	14.2	15.3
25–44	25.0	23.1	22.9	23.2	17.9
45–64	25.1	25.0	26.2	24.2	21.7
65 and over	12.4	14.9	15.2	14.3	13.5
Total	100.0	100.0	100.0	100.0	100.0
Dependent groups	35.4	39.4	37.3	38.4	45.1
Active groups	64.6	60.6	62.7	61.6	54.9

Source: Sample Census, 1966.

The rural communities have more in the dependent and fewer in the active age groups than the overall national pattern. The continuance of this trend has obvious implications for the future viability of such communities. Population decline also has the effect of changing the sex ratio, that is females per 100 males, particularly in the young unmarried age groups. The sex ratios in the single active age groups are compared in Table 10.8.

There are fewer females than males in the 15–44 single age groups for the population as a whole, but the rural communities, especially North Norfolk, show a much greater imbalance in all age groups. The statistics for the Scottish and Mid-Wales areas conceal high inter-county variations. For

Table 10.8

Sex Ratio of the Single Population (15–44) in Selected Areas, 1966

Age group	England and Wales	Highlands and Islands	Mid-Wales	North Norfolk	North Mayo
15–19	90.7	86.4	93.3	56.8	86.0
20–4	61.4	62.4	54.9	48.1	55.8
25–9	50.2	54.1	46.6	40.0	48.9
30–4	58.8	63.7	49.3	30.0	35.2
35–9	67.2	63.3	42.1	42.9	39.3
40–4	80.1	91.9	64.7	36.4	40.1
Total (15–44)	75.0	73.7	68.9	49.7	62.9

Source: Sample Census, 1966.

example, the county of Sutherland has sex ratios of 46.1 and 42.9 in the age groups 20–4 and 25–9 respectively, compared with figures of 62.4 and 54.1 for the Highland area as a whole. Similarly, the Mid-Wales area statistics conceal inter-county variations, especially in Radnor and Brecon.

The vicious circle theory

The foregoing has illustrated the effects of depopulation on the age structures and sex ratios of the communities under study. Perhaps the most serious problem of all is the cumulative and circular aspect of the depopulation process. It has been suggested earlier that lack of employment is one of the main causes of population decline. As a result of the decrease in population fewer services are demanded and these in turn contract. For this reason the economic attractiveness of the area in question is reduced. The end-product is that employment is reduced further and the vicious circle is complete.[3] Similarly, when the age structure or sex ratio is unbalanced, the community in question becomes known as a residual one and the imbalance becomes even more pronounced as a result of further selective migration. Factors that were once effects soon become causes. Eventually, for economic reasons, even local government authorities find it necessary to curtail their commitment to such areas in the form of water supplies, sewage disposal, road improvement and public services in general.

Forced migration

One last point can be made in relation to the effects of population decline. Few would argue that some migration within the confines of a country is inevitable, but such a proposition may not be acceptable if a large proportion of the migration is involuntary. In the USA, Malzberg and Lee refer to the disturbing effects of unfamiliar communities and social situations and suggest that mental illness, crime, political instability and other forms of social disorganisation are positively associated with geographical mobility.[4] There is no evidence of a comparable nature available in the UK, but the effect is not inconceivable. In the North Norfolk school survey already mentioned 179 respondents (45.7 per cent) wished to remain in the area, 121 (30.8 per cent) were undecided, while only 92 (23.5 per cent) wished to leave. It is fairly certain that a large proportion of the 179 respondents wishing to remain will have to migrate due to a lack of suitable employment opportunities. If the American findings have any relevance here, the human problems involved in involuntary migration may be quite serious.

Conclusion

A common characteristic of the areas examined is the decline in total population and the migration of the younger age groups to work outside the area. A major cause of this movement has been the reduction in agricultural employment, arising from changing technology and economic pressures. A second cause has been the chronically lower level of wages and earnings in agriculture compared with industry acting as a 'pull' factor.

Among the effects on the areas have been unbalanced age structures and sex ratios in the rural population, which tend to label them as problem areas. A high migration rate leads to a decline in demand for goods and services, which in turn disadvantages the area even more. This leads to an apparent lack of commitment on the part of both local authorities and private interests towards social provision.

The impact of the EEC on these areas will be in direct relationship with the changes, if any, forced upon their agriculture. If these changes are deemed to be politically unacceptable then remedial action must be taken, either to support the local agriculture or to provide alternative employment in the area. At present it would seem that the current trends will continue unless active steps are taken to stop or reverse them.

References

1) Robinson, D., 'Low-Paid Workers and Incomes Policy', *Bulletin of the Oxford Institute of Economics and Statistics*, XXIX, 1 (1967).
2) Bracey, H. E., *Social Provision in Wiltshire* (Methuen, London, 1952).
3) See Myrdal, Gunnar, *Economic Theory and Underdeveloped Regions* (Gerald Duckworth, London, 1958), pp. 11–22, for a discussion on the theory.
4) Malzberg, B. and Lee, E. S., *Migration and Mental Disease* (Social Science Research Council, New York, 1956).

11 The Economic Potential of Rural Areas

Professor M. GASKIN
Department of political economy,
University of Aberdeen

The operative phrase of the title of this paper, 'economic potential', is not a precise term. One might substitute 'possibilities' for 'potential', but while this conveys a more neutral tone, it is equally vague. To find some bearings on the matter one has to look at the contexts in which the term is most used, and here one can point to two kinds of questions in regional studies in which it is frequently pressed into service.

One sort of question is asked from what, with some strain on meaning, can be called a 'positive' standpoint. It calls for a consideration of the kinds and quantities of economic goods and services that particular collections of resources – in this case, rural resources – could produce over some future period. Such an exercise involves an examination of such factors as the qualities and quantities of the resources and the nature of the demands likely to be placed upon them. This approach is labelled 'positive' on the ground that it invites one to proceed on as factual a basis as possible, without preconditions about the uses to which rural resources should be put, or the kinds and levels of activities that should be pursued within rural regions. It cannot exclude all preconditions; any system of resource use is subject to constraints, particularly those imposed by the institutions through which they are controlled. Also, this approach is not 'positive' in the precise, philosophical sense of the term, since implicitly it accords priority to output as an index of 'potential', and our measures of output rest ultimately on questions of value – subjective, not monetary – and not of fact.[1]

There is another usage of 'economic potential', frequently met with in regional studies, where the standpoint of the questioner might conveniently be described as 'normative', in that his interests are linked closely with some specific policy objective for the area in question. Thus to ask what the economic potential of an area is, is frequently to consider what viable economic activities can be imported in to, or promoted within, the area in order to achieve some specific objective. In the rural case the objective, whether stated or not, almost always has to do with population – maintaining a particular level of population or altering its rate of change, with or without attached conditions about the settlement pattern to be preserved or achieved.

These two approaches are not so different in kind as the terminology suggests; in a sense, it comes down to the measure of 'economic potential' adopted – in the one case output, in the other employment. In this paper both standpoints are adopted successively, beginning with a broad factual look at rural resources and the potential claims upon them, and offering some comments on the problems of their optimum use. Then, on the normative tack, the question emphasised by Mr Wallace and Mr Drudy, of whether and how one may provide an economic base for sustaining population in rural areas, will be taken up.

Rural resources and their potential

Put simply, and in a neutral way, the economic potential of rural areas resides in the resources that lie within them and the way in which these resources are used. The most copious and ubiquitious resource is land; indeed definitions of 'rural areas' sometimes include the predominance of activities making extensive uses of land as a prime criterion of rurality.[2] The 'land' referred to here is not simply the soil; it must be defined, as economists have always done, to embrace everything that is naturally connected with any particular parcel of it – its fertility, climate, mineral content and situation. Indeed when one talks of rural land these days one must stress a whole array of qualitative attributes that are geographically or historically determined. For example, some rural areas have aesthetic qualities born partly of natural elements like soil and climate, but especially of the human care and inspiration that have been lavished upon them and which, designedly or otherwise, have invested them with high qualities of scenic beauty. Although land is the most obvious and widespread resource of rural regions it is not the only one. Labour and invested capital also exist in rural areas and questions connected with their supply and efficient use are very relevant to rural potential. Paradoxically, some major problems of rural areas (in the normative realm) turn on what is relatively their scarcest resource, labour.

The rural labour force

The size of the rural labour force cannot be assessed with any precision. Clearly it will depend on how one defines rural areas, and where the boundaries are drawn is an arbitrary matter. Recently, R. J. Green,[3] using a fairly broad criterion of settlement size and pattern, has divided the UK into five 'conurban' and six rural regions. In the rural regions which he defines for England and Wales he estimates a population of four and a half millions living in settlements of smaller than 7,000 people, with slightly more than this number inhabiting 139 larger places. For Scotland, outside the central

142

conurban belt, one could add some 850,000 people living in rural settlements of under 7,000, with another 600,000 in 20 larger places. These are figures of population, and not of labour supply. However if we make the rough assumption that Green's four and a half millions living in 'small' settlements, plus the corresponding Scottish figure, constitute the population from which the rural labour force is drawn, we might estimate the working population at between 44 and 47 per cent of this total; that is between two and two and a half millions. But this seems altogether too high and an alternative calculation can be made that starts at the other end, with the totals of those engaged in primary industries.

In mid-1971 the Department of Employment's figure of employees in primary industry excluding mining was 350,000, or 1.1 per cent of the national total. As an estimate of rural labour resources this is no more than a beginning and requires some major adjustments. To begin with the numbers of self-employed people in primary industries must be added, and a minor subtraction made for fishermen living in such non-rural settlements as Hull and Grimsby. These adjustments cannot be made with precision, but an overall total of 600,000 may indicate the rough order of the rural work force, employees and self-employed, in primary industry. However, yet more adjustment is needed to arrive at the total rural labour force. For example, part of the working population statistically attributed to other industries is engaged within rural areas in providing services that are essentially integral processes of the primary industries themselves; transport is the obvious one, and probably the most important. Yet another part, of greater dimensions, is employed in jobs that are sustained from the general expenditures of those working in primary industry. Most of this employment is in the service trades, including the substantial component engaged in providing public services which, while not 'purchased' directly, are provided in amounts related to the size of the rural population. One can probably double the figure of 600,000, to arrive at a total of primary plus primary-based employment and rough though it is, this is a more likely estimate of rural labour than the much larger figure based on Green's totals of rural population.[4]

The question to which estimates of this kind most directly lead is that of the availability of labour within rural areas for non-primary activities. To anticipate later discussion, since the decline of primary industries as employers is a prime fact of the rural situation it is natural to ask what alternative activities could be promoted in these areas to provide a new basis for the economy, and to realise the potential of the resources at present there. Any selection of these must inevitably include manufacturing industry, and the immediately relevant point is that such industry is primarily labour- and capital-using, rather than land-using; its space demands may not be negligible but labour is all-important.

Now, at one million or over, the primary-based labour force of the UK is not negligible. But caution is called for in moving from this to an estimate of potential labour 'reserves' in rural areas. One must, for example, deduct most of those engaged in trades and occupations supplying services to the 'base' population of the rural areas, since many of these people would be needed to serve the same population however else engaged and where ever else located. Secondly, the present decline in employment in traditional rural industries must surely decelerate and bottom out in the not-too-distant future, though there are no clear signs of this yet. Granted that some lower limit will be found, it must be regarded as a quantity of labour that will be unavailable for alternative employment. Finally, one should also bear in mind the age composition of the rural labour force; this is almost certainly less favourable than the national average from the point of view of an efficient labour supply.[5] The upshot of these qualifications is to point to a total considerably below one million as an available labour supply in rural areas. But even at half this figure it would, on the national view, constitute a substantial reserve. Given the tightness of the British labour market – the present recession apart – and the inhibiting effect that this has probably had on economic growth, a labour reserve of this scale merits serious consideration in the framing of policy. It is important too that this segment of the nation's labour force, like any other, should be employed in activities that give it its highest value in use. The great drawback to this prescription, of course, is its sheer dispersion, the fact that it is scattered and unavailable at any point in amounts that much modern industry requires. This considerably reduces its value *in situ*, though it points to the question whether, suitably redistributed, it could still be employed within the rural regions. To this we return later.

Capital resources in rural areas

In the case of invested capital the qualifying participle is all-important. Capital in an uninvested or unembodied, that is financial, form is mobile and unattached to particular geographical locations. But past investment has resulted in a mass of capital works and improvements of the rural environment that comprise important resources for use in the present and the future. A great deal of this rural capital is combined with land, either for agriculture or forestry, in the form of drainage and other improvements. Indeed the union of the two resources is frequently so close that for many purposes distinctions are pointless. But much rural capital is in the form of houses, schools, churches and other public buildings, as well as in systems of public roads, drainage and lighting – everything, in fact, embraced in the term 'infrastructure'.

144

It is a commonplace of regional analysis that the existing infrastructure of an area may constitute a very significant resource when appraising its 'economic potential', but in the rural context the matter is not at all straight forward. For example, it is relevant to ask whether infrastructure contains an element analogous to the 'going concern' component in the valuation of a business. A community, whether a village or a small town or even a scattered rural parish, may be economically, as well as socially, more than the sum of its parts. It is a society making use of a group of real assets that have been assembled and co-ordinated in particular ways, over time. The potential of the total system as a going concern may well be more than that of the individual items and sub-systems that compose it. If this proposition is valid it would imply that the creation of new communities or the significant expansion of old ones might cost more than the replacement cost of the assets of the older communities. This amounts to saying that new communities may be less efficient than older ones in the sense of yielding lower levels of service for a given outlay of resources.

But the matter is complicated by considerations governing the continued use of existing assets in rural areas. Whether or not it is economically advantageous to do this is not a clear-out issue. Most real assets depreciate in the two senses of suffering physical deterioration and becoming obsolete. In the long run, most investment is temporary and the question of whether to replace assets in one location rather than another may arise. In the present context the significance of this is enhanced by the fact that, like much of the social capital of the UK, many items of capital in rural areas are old and could well do with being replaced. But further, assets that to the ordinary eye have a lot of life in them become uneconomic to operate when the total costs of running new, replacement assets fall below the operating, or prime, costs of the old; economically speaking, the old assets are then obsolete. In the case of certain elements of social capital, housing for example, cost comparisons of this kind are not a clear-cut matter; and wherever the existence or scale of a community is involved, as is so often the case, imponderable social values complicate the issue. Nevertheless, the economic considerations are an important factor in the situation, and have a bearing on the problem raised earlier of the usability of rural labour resources. Briefly, at this point, one important action in the face of an inconveniently dispersed population is to concentrate it more – to select certain settlements for growth and put the new assets which are constantly needed in replacement of those wearing out, in these places. Indeed a full evaluation of the economic situation in rural areas, one that took account of the prospects for rural industry of all kinds, might lead to the conclusion that such concentration should not wait on the progressive depreciation of assets, but proceed at a rate that would involve the abandonment of many otherwise viable rural assets, perhaps in whole communities. If this is the case the inference is that, on a full view, the potential of many existing assets in the rural areas may be very much less than appears to the inspecting eye.

However, today the countryside is beset by very diverse developments and the situation is not one for easy generalisations. Two such developments have an important bearing on major rural assets like houses and communal services. One is the widening range of commuting, with the concomitant spread of occupancy of rural housing by people working in urban areas. This has already changed the character of the once rural villages that lie nearest to the conurbations. As roads improve and the desire of many town dwellers to escape from an urban environment strengthens, this movement will push ever deeper into the countryside, absorbing and finding a use for some of those three in four rural villages for which Professor Wibberley foresees no continuing rural purpose.[6] Furthermore, it has some powerful social and economic arguments on its side, the economic ones resting on the saving of cost over the construction of completely new communities. The second development is the spread of the ownership of second homes by town-dwellers. This trend, now firmly established, spreads much more deeply into the rural areas, reaching some of the remotest. One striking effect of it is that, like all such urban intrusions, it is creating a new perspective on 'obsolescence' in rural assets. The second-home owner is frequently prepared to acquire dwellings the fabric and location of which have become unacceptable to rural workers. In this sense the potential of these rural assets has been increased and may continue to rise. It is arguable that second homes help to sustain the viability of rural communities. There is no doubt that they preserve some employment and income that would otherwise drain away from these areas, although in view of the elements of subsidy present it is difficult to assess how far this contribution is economic from the national viewpoint.[7] This notwithstanding, the second-home trend is merely accelerated, not created, by subsidies, and the prospect is that it will continue. It is undoubtedly revalorising, as well as restoring, many rural assets that had seemed bound for economic and physical oblivion.

Land resources

It is clear that the most ubiquitous resource of rural areas, land, is coming, and will continue to come, under increasing pressure. Demand for land for one purpose or another, in many places for various competing purposes, is increasing and shows no likelihood of diminishing. The causes are not far to seek. The population of Britain is growing not rapidly but steadily, and will go on doing so possibly for another two decades. The latest estimates of the Registrar General forecast a rise of about nine millions by the end of the century; while this will almost certainly prove to be an over-estimate, the tide of population will assuredly reach higher levels yet. Furthermore, this population will be richer and more mobile. It will want more living space and many of its members will be able and prepared to buy it, if allowed. Numerous others will, as a matter of social policy, have to be provided with more space as our crowded industrial regions are redeveloped.

The greater personal mobility of the population, conjoined with greater affluence and more leisure, will transmit the pressures of recreational demands on land far afield in the countryside. Very few recreational needs are large, exclusive users of land, but individual activities put demands of some kind on large areas – for example, for access – and in doing so create constraints or conflicts over land use.

Other uses of rural land are planned to increase, notably water catchment and forestry. The rising *per capita* consumption of water by an increasing population demands, and is now beginning to get, a reorganisation of the water industry and a comprehensive appraisal of the country's water resources Overall the demands of water supply are unlikely to constitute a big user of land, but they will impose sharp pressures on particular places, mainly, though not exclusively, in upland areas. Forestry, under public and private ownership, has been since 1945 a much more extensive competitor for land. Under targets laid down in 1967 the Forestry Commission's programme alone implied a rate of planting which, if achieved and maintained for the rest of the century, would increase its afforested area by two and a half million acres, or about 3 per cent of the total land area of the country. Following the recent review of forestry policy, a lower rate of new planting is envisaged, one that will involve a tapering off of planting on unafforested land as the acreage of replantable land rises with the felling of timber now growing.[8] However, for some time yet land will continue to be transferred to forestry from other uses and this may impinge sharply on some areas, again in the uplands.

Implicit in discussions of this kind is the assumption that as the 'residual user' of land, agriculture will be the only significant sufferer from the transfer of land to these other uses. This view contains a core of truth, but it should not obscure the real, and locally very important, competition between non-agricultural users of land. Furthermore, the prospect of EEC membership begins to throw doubt on the identity of the 'residual user' of land in some areas. But before looking more closely at the possible evolution of land use, some quantitative perspective is called for.

Any attempt at a quantified forward look at land use in the decades ahead must inevitably lean heavily on the work of Professor Wibberley and his co-workers at Wye College. With the publication (in 1971) of Edwards and Wibberley's *An Agricultural Land Budget for Britain, 1965–2000*, we now have a carefully argued and succinct summary of their collective prognoses for land use during the remainder of this century. Three major conclusions that emerge, put in the briefest terms, are: that space demands for all urban purposes up to the year 2000 are likely to raise the urban area of Britain from eight and a half to 11 per cent of its land surface; that forestry will by then occupy about 11 per cent of that area; and that the losses of agricultural output due to these reallocations of land are likely on foreseeable trends to be more than made good by the increase in agricultural productivity. Mrs Edwards and Professor Wibberley arrive at their estimates on the basis of various assumptions about rates of growth of population, densities of urban development, degrees of self-sufficiency in agricultural output as dictated by policy, and rates of growth of agricultural productivity; and they examine the effects of alternative values of these variables. Naturally such assumptions are surrounded with uncertainty; and while the effects of some variations may not affect their results greatly some possibilities may be noted. Future population growth is a major uncertainty. Already the Registrar General has revised his forecast for the year 2000 to a level below the lowest assumption of Edwards and Wibberley; and who can doubt that it will not come down more yet? On the other hand, one can be optimistic enough to expect rates of growth of *per capita* income at, or above, the 3 per cent upper limit of their assumed range. However, these two revisions would work in opposite directions in their influence on residential and recreational demand for land, so that unless there were dramatic variations they would not materially affect the broad conclusions of the study.

The demand for land for forestry is now certain to alter, and to fall below the projections of the Wye study. In consequence of the recent review of forestry policy – a review that was accompanied by an extensive cost-benefit study of the forestry enterprise in Britain – it is now proposed that the 1967 planting targets for the Forestry Commission, which underlay Edwards and Wibberley's estimate, be discarded.[9] Instead of a projected rise to an annual 60,000 acres of new planting by 1976, with a prospect of 70,000 acres per annum thereafter, the government now envisages 'a combined planting and replanting programme of up to 50,000 acres a year'. This objective implies a declining rate of absorption of new land by the Commission as trees, mature and land is released for replanting. By 1990 such restocking would be running at 40,000 acres per annum and, by the end of the century, at 60,000. Furthermore, changes in the system of grant aid to private planting are proposed that would have a broadly similar effect of maintaining the rate of planting and replanting combined, at least in the remoter areas, with an implied decline in the planting of new land.[10]

The future of British agriculture in the EEC raises another uncertainty about the character and relative intensities of differing demands for land in the future. Edwards and Wibberley, writing when they did, took no account of this; they examined the implications of alternative levels of self-sufficiency in food production on the assumption that balance of payments considerations would at least make this an object of concern for future policy.[11] How this position will be affected by membership of the EEC is difficult to forecast, but it may not greatly affect the picture. The more important factor over the final quarter of the century may be the world food balance as a whole. At the moment it requires some optimism about population policy and agricultural productivity to predict a declining requirement of food production from British land.

Land is not homogeneous and this prognosis is consistent with a falling demand for some types of land. As far as agricultural claims on land are concerned, it is in the upland areas that the viability of particular farm enterprises in the UK is most in doubt. It is true that Dr Sicco Mansholt, speaking in Scotland early in 1972, gave a verbal assurance that the assistance at present extended to hill farming, and in present market conditions essential to its survival, would be consistent with the CAP. It appears that, so far as a Community approach to regional policy has begun to take shape, some level of subsidisation would be accepted on these grounds also. But two clouds mar this particular horizon. First, a primary object of the Mansholt plan was to bring about a reform of farm structure. While this is directed above all at the fragmented, small-scale structure of European peasant agriculture, when adopted (this is hardly round the corner) it is bound to have some effect on British farming, particularly the smaller-scale farming of the upland areas. Secondly, and quite apart from any future effects of the CAP, many pressures, including the rising opportunity cost of labour referred to by Wallace and Drudy, are combining to reduce the numbers of small farmers, again most markedly in the upland and marginal areas. This will maintain, if not accelerate, the decline of the agricultural population in these areas.

In the Mansholt plan it was envisaged that the combination of farm restructuring and the increased productivity of the remaining agriculture would lead to some considerable areas of land passing out of agricultural use. The suggestion is that this land will be used for alternative purposes such as forestry and recreation, though the economics of this raise many questions. Assuming that eventually we see this development in Britain, what will be the likely response to a situation in which the traditional 'residual user' of land withdraws from this role? The obvious thing to say is that we should not assume that land that becomes uneconomic for commercial agriculture will therefore be unwanted for other private purposes. Some upland and marginal land released from farming will be acquired for private sporting purposes, a land use in flourishing condition at the present time; or it will be wanted as an asset that is considered to have a durable value in real terms (like gold, where thinking makes it so), or simply for the subjective satisfaction that land ownership has always brought to many. Where such land is in easy travelling distance of urban areas we should not exclude a possible spread of part-time farming, with labour and other inputs uncosted; shortening hours of work in non-rural industry will make this increasingly possible, and the cultural environment of an increasingly affluent society may enhance its appeal. Presumably EEC policy would attempt to discourage this; but it might be difficult to do so unless rigorous, and possibly unacceptable, controls over land tenure were enforced.

Maximising 'potential'

The economic potential of rural resources is partly a function of the purposes and methods of their use. They can be allocated more or less efficiently to meet the demands that society places upon them. How one might approach this at the social level – the quest for modes of optimising the use of rural resources – is a large question that cannot be opened up here, but this section of the paper cannot be closed without a remark on the matter, particularly as it affects land use. More than any other resource, land is subject to a complex pattern of competition and complementarity in use. Almost all uses of land are competitive with one another, but some serve the ends of more than the prime user, as when forestry confers benefits on neighbouring agriculture, or a reservoir opens up new recreational possibilities to a nearby population. At the same time, the use of land may involve sharp and widely extending conflicts of interest, conflicts that endlessly complicate decisions in this area. In some cases cost-benefit analysis[12] and similar optimising techniques can help to clarify, by partly quantifying, the bases for decision. In the present state of the art these techniques are subject to well-known limitations, which apply with particular force in some problems of land use, especially those involving widely diffused interests. Nevertheless, the approach to land use, and to rural resource use in general, must start from an economic base; that is, from one that pays regard to market valuations – of inputs and outputs – as prime criteria for decisions. Too often in the past, and still today, statements are made about land use that assume that physical qualities – soil, configuration, climate and so forth – should be the determining elements. But decisions aimed, however imperfectly, at maximising the 'economic potential' of rural resources must have regard to wider criteria. In many, perhaps most, land use problems the economist's approach by itself can take one only part of the way, but properly used it offers a firmer basis for grappling with the imponderable factors in resource use decisions than approaches grounded solely in the applied sciences.

150

Sustaining rural areas

Turning now to a 'normative' interpretation of 'potential', what possibilities offer themselves for maintaining population levels in rural areas? As was pointed out by Wallace and Drudy, rural population has declined partly because of changing technology in traditional industries, but partly also for reasons of a social rather than an economic kind. How may this trend be arrested, assuming that this is a policy objective determined by a political authority?

We may begin by asking in what way the level of population in the countryside is related to rural resource use. As we have seen, some important approaches to resource use emphasise an output objective, particularly output maximisation. In certain cases, notably in some ecological approaches, maintainability of output is made the prime objective, but this qualifies rather than contradicts the proposition. On the other hand, the problem of rural population *prima facie* requires one to consider employment as the maximand rather than output. In practice, the conflict between these two objectives is less than it may seem. Depopulation of rural areas – of any area for that matter – is not simply a matter of job opportunities. Income possibilities are also crucial; if rural-urban income differentials widen too much, the countryside will fail to hold its people regardless of the existence of jobs. Hence in rural population policy one cannot simply aim at an employment target without regard to the outputs, and hence the incomes, attached to that employment.

Under a variety of pressures and incentives the traditional land-using activities of the rural economy have raised their productivity, per man and per acre, with signal success, so much so that they are a major cause of the problem and cannot contribute to its solution. With the exception of the residential use of rural land, and in a more limited way forestry, none of the land-using activities discussed earlier can do much to maintain population in the countryside. The residential use of rural areas, in which is included the absorption into use by urban workers of rural communities as commuting ranges extend, quite directly sustains population in some rural areas, simply by locating people there. It also creates some secondary local employment in the service trades. This development is not one that the dedicated conservers of rural life regard with favour, but it is a fact of life that is with us and will continue so. Furthermore, verdicts on its social effects rest on value judgements about which there is no general consensus of view.

However, the influence of rural residence has its bounds. Large rural tracts will continue to lie outside commuting range and for these other means of sustaining jobs and population must be found if specified levels of population are to be retained within them. Among the other extensive users of land, forestry, or the timber industry in general, has a contribution to make to the rural economy and this will be quite significant in some limited areas. Forestry operations as such are not large users of labour; in March 1971 the total employment of the Forestry Commission, which owns half the country's forests, was 9,515, not all of whom were located in rural areas. If one adds employment in private-sector forestry the figure rises to about 25,000, but the future trend of this total is problematical. The Forestry Commission's own labour force is declining – it fell by 5 per cent in 1970/1 – and if one takes standards of labour productivity in the Scandinavian and Canadian industries as a guide this trend has far to go. However, in some areas the trend will be offset in the Commission's operations, and probably in the private sector also, as forests now growing move into a more labour-using phase of their cycle.[13] If one brings in the possibilities offered by the wood-using industries the picture brightens still further. As timber presently growing matures, at least two new pulp mills may become feasible in Scotland, and there are possibilities for extending the manufacture of chipboard and similar timber-based products. One recent commentator has put the possible number of new jobs created in forestry and related industries in the Scottish Highland region alone at 4,000 by 1985.[14] This is a very substantial figure in the context of that area and would contribute appreciably to maintaining population in an economically difficult region. If a similar proportionate level of growth were possible in other regions, then the role of forestry in sustaining population, in the remoter rural areas at least, would be far from negligible. But this, clearly, is not on the cards.

In the aggregate, set against a rural labour force of rather more than a million and a 'reserve' of perhaps half that figure, the total contribution of employment in the timber industry, estimated at 36,000,[15] is not great. The recent policy statement outlined a future policy for forestry in which emphasis would be placed on 'wider social objectives', notably creating and maintaining employment, creating and preserving amenity and fostering the recreational use of forests.[16] However, where the paper gives some specific indications of a future programme bearing on employment, in its comments on the future operation of the Commission, the actual implications for employment are less than clear. For the Forestry Commission the paper proposes a 'combined planting and replanting.... of up to 55,000 acres a year [which] would broadly maintain the Commission's contribution to employment in the key areas of Scotland, Wales and the north of England'[17]. It is not clear whether this relates solely to employment in 'planting and replanting', or to total employment in forestry, and if the latter whether it takes account of the potential and complex effects on employment of rising productivity in timber extraction coupled with an increasing harvest of

152

mature trees. As already noted, the policy statement also proposes changes in the grant aid to private forestry; these would include the ending of the dedication scheme, but with the possible retention of a grant 'designed to encourage private planting to follow the wider social objectives proposed for State forestry...', including the maintenance of employment in certain rural areas.[18] The potential effect of the new policy would seem to strengthen the conclusion of the commentator referred to earlier,[19] that in regard to forestry operations in the Scottish Highlands any significant increases in employment in timber industries over the next three decades will come from the exploitation of existing forests, not from the planting of new ones. As far as land use is concerned, it has been noted already that the policy statement implies a diminishing demand for new land on the part both of the Forestry Commission and the private sector.

The recreational use of the countryside, important though it is and will continue to be, may prove a weak reed on which to rely for rural employment. Of course 'recreation' covers a variety of activities from day trippers motoring to local beauty spots to the long-stay tourist taking an extended holiday, mobile or static, in the remoter countryside. All create some income and some employment in the places they visit. Some developments, for example the country park offering a wide range of attractions from wild animals to funfairs, may create quite extensive local employment, but two points are noteworthy about such parks. One is that to operate on a large-employing scale they must be within reasonable travelling distance of large urban populations. Secondly, as with most recreational activities, the employment they offer is seasonal and unless it can be synchronised with other rural activities the jobs created will tend to be filled by migrant workers. Country parks in all their varieties have an immense part to play in exploiting the recreational potential of rural areas, as well as in channelling these activities in directions that minimise conflict with other rural interests; but neither they, nor tourism generally, can make more than a modest contribution to sustaining rural population by the creation of employment.

One is left then with essentially non-rural activities, non-extensive users of land, as the main hope for retarding the decline of rural employment. The most important of these is manufacturing industry, with perhaps some contribution from service activities serving non-local markets. The prospects for attracting manufacturing industry to the rural areas are conditioned by two broad sets of influences. One is the amount of new industry, or old industry on the move, which is 'there' to be attracted. The second embraces those factors intrinsic to the rural areas that affect their ability actually to capture new industries; these factors are what are usually comprehended in 'economic potential' when the term is used in this context.

On the first question there is a rather pessimistic view abroad at the present time, a view for which there are some compelling grounds. If, looking at post-war experience in stimulating new industry in the lagging regions and taking account of the more extensive regional policies of recent years, one makes a guess at the number of 'mobile' jobs likely to become available over, say, the next decade, and then sets this against existing commitments to provide employment in new and expanded towns, one can have doubts as to whether the number of jobs available will match up to the needs of regional policy as a whole. Within the broad concerns of regional policy the rural areas do not, for understandable reasons, rank very high.

Membership of the EEC offers one ray of light here in that it may lead to an influx of overseas firms, particularly American. These, being by their nature initially mobile and generally more resistant than domestic industry to the attractions of the prosperous regions, will add to the total of new industry available for diversion to the rural areas. But plants of this kind are frequently, though not invariably, of medium to large size in employment terms and this inevitably draws them to urban sites with access to adequate labour supplies.

Here one arrives at those factors that control the ability of rural areas themselves to attract new manufacturing industry. There are two main ones that call for consideration: transport costs and labour availability. There are strong grounds for doubting that transport costs are the obstacle to the rural location of manufacturing industry that they are frequently held to be. From the evidence revealed in studies of location and industrial mobility [20] it appears that for many industries, even in a comparatively distant region like Scotland, transport costs account for a very minor fraction of total costs. Of course there is the danger, when looking at the relative scale of transport costs of existing industries in any area, that the sample is biased by the mere fact of survival, and the contrary view has been argued on this and other grounds. [21] Nevertheless, the bias argument itself can be overplayed since the very fact that firms in a range of industries can operate successfully in non-centrally situated areas is evidence for the ability of firms to cope with transport costs in these cases. [22] The results of the studies referred to, as well as other evidence, [23] show that there are sufficient industries for which relative transport costs lie within the normal range of variation of total costs between firms, or else within the limits of market price differences made possible by qualitative differences of product. This view is strengthened by the reflection that a number of extensive rural areas – East Anglia, the South-West, Mid-Wales – are hardly far flung in relation to major zones of industry and population. It is true that for the really peripheral regions the list of possible industries will be smaller; but even there the transport factor is not decisively adverse, especially where differentiating features of quality can

154

be exploited. What the economist calls 'imperfect competition' is the friend of the remoter areas.[24]

Some factors connected with distance and remoteness may constitute deterrents to industries moving to rural locations. These include the time factor in the transport of goods and materials and in the travel of personnel as well as physical remoteness from customers, suppliers or parent plants, unmitigated by access to rapid transport facilities. These 'distance costs', as they are frequently called, rather than the simple financial costs of moving goods, may have a strong deterrent effect. For the very remotest locations, inhibitions of an almost psychological kind, especially as they affect the managers and key workers who usually have to be brought in, may be strongly operative.

There are undeniable differences of opinion about the significance of the transport factor in rural development. All discussion of the point is complicated, if not vitiated, by the fact that in rural areas distance is rarely the sole adverse element in the situation. Indeed, almost invariably there is present a more potent deterrent in the form of a small and restricted labour market. We have seen earlier that the potential labour reserve in the rural areas is not negligible *in toto*, but when viewed in relation to the area over which it is spread it is revealed as very dispersed and thereby greatly reduced in potential for new industry. There are areas where the rural population is relatively dense, especially where numbers of settlements lie grouped within easy travelling distances.[25] Furthermore, increasing mobility is undoubtedly widening the labour catchment zones round any single rural location. But as a general rule the UK lacks the dense occupation of the open country which in some European countries, notably Western Germany, has favoured the industrial development of towns in agricultural areas.[26] At the same time, the conditions of labour shortage that impelled German industry to move into such areas have existed – the current recession apart – in Britain, and may well continue in the future. If the labour progressively released from agriculture and other rural uses could be concentrated in larger settlements to form bigger and hence more attractive labour markets, some rural areas could acquire the potential to support small to medium-scale industrial plants. The difficulty remains that they may be competing with the older industrial regions for what proves to be an insufficient amount of mobile industry. The handicap with which the rural areas start in this race can be moderated if, by positive planning measures involving the location of housing as much as of industry itself, development can be concentrated in selected zones. In this way the positive side of the picture can emerge, particularly the high amenity that many rural areas can offer to industrialists and to the staff they bring in. Furthermore, provided the right centres are chosen for development, advantage can be taken of the community services and infrastructure that already exist in these places, thus realising the potential of some of the capital assets in the country towns. Finally a contribution can be made to the pressing problems of congestion in the conurban regions.

There is one class of industry that by its nature is frequently compelled to locate in rural areas and for which special provision of infrastructure and housing is usually necessary in order positively to assemble a labour force. The industries concerned are those based on the extraction of natural resources or the exploitation of particular sites for processing purposes. The location of two of three new aluminium smelters, one in Anglesey and the other at Invergordon on the Cromarty Firth, are prime examples in recent years, but they were preceded by others, for example, the pulp mill at Fort William, as well as earlier aluminium smelters in the Scottish Highlands. Now, with the discovery of oil and gas in the North Sea, these developments are being matched by others, some of them even larger in employment terms. The full impact of these recent plants on their localities has yet to be seen, and it will be some time before one can measure it fully, especially where there will be a heavy concentration as on the Cromarty Firth. However, already such developments have demonstrated that what, in rural terms, are very substantial labour forces can be mobilised when there is a strongly impelling motive on the part of the industrialist and a proper backing from all the interested authorities.[27]

Undoubtedly these examples are exceptional. Without the compelling need to be near a mineral resource, or alongside deep water, industrialists are not easily persuaded that rural locations are practicable and offer distinct advantages in some respects over sites in industrialised or congested regions. Indeed, these resource-based industries apart, there will be difficulty in sustaining the present level of population in rural areas by the stimulation of rural manufacturing employment alone. At least as much will be achieved by the increasing use of many rural areas for residential purposes. Together these two potential supports of rural population will produce a very different society; one much less rooted the countryside though, paradoxically, it may be the search for 'roots' that sends it there.

References

1) Graaf, J. de V., *Theoretical Welfare Economics* (Cambridge, 1957), p. 1.
2) Wibberley, G. P., *Rural Activities and Rural Settlement*, TPA Conference (1972), p. 1.
3) Green, R. J., *Country Planning* (Manchester, 1971), pp. 4–5.
4) An inspection of the maps to Vol. III of the Redcliffe Maud Report (Cmnd. 4040–II) shows that within Green's rural regions there is quite a moderate amount of commuting to urban settlements. As some of these settlements are quite large, it is clear that a proportion of his rural population is employed in non-rural industries.
5) Though, arguably, older workers are not always less efficient – or, at least, less desirable from an employers' point of view. *Forestry in Great Britain, An Interdepartmental Cost/Benefit Study* (HMSO, 1972), p. 11.

6) Wibberley, G. P., *Rural Conservation*, RICS Conference 1970, Paper 1, p. 4.

7) This movement is, of course, encouraged by certain elements of present housing policy and local government finance. For example, many second homes are being rendered habitable with the aid of grants, the original purpose of which was to improve the standard of first homes.

8) *Forestry Policy* (HMSO, 1972, paras 35–37). The precise status of this paper is not completely clear. It is presented as a governmental 'review' of policy, and one that has been 'greatly assisted' by the Treasury's cost-benefit study of the whole forestry enterprise. However, the Foreword states that the 'views of interested bodies will be welcomed', which rather puts it into the Green Paper class, in spite of its white covers. The cost-benefit study, published separately, and noted above (reference 5), was a very comprehensive exercise that essayed the difficult task of evaluating the amenity and recreational benefits of forestry.

9) *Ibid., loc. cit.*

10) *Ibid.*, para. 41.

11) All their assumed values lay above the 60 per cent level of self-sufficiency in temperate foodstuffs that obtained in 1965. They considered that future levels would lie within the 65 to 75 per cent range.

12) On the application of cost-benefit techniques to land-use problems see the comprehensive survey of recent applications by Professor G. H. Peters, 'Land Use Studies in Britain: A Review of the Literature with Special Reference to Applications of Cost-Benefit Analysis', *JAE, XXI*, 2 (1970), pp. 171–214.

13) Hampson, S. F., 'Highland Forestry: An Evaluation', *JAE, XXII*, 1 (1972), p. 54.

14) *Ibid., loc. cit.*

15) *Forestry Policy*, para. 17.

16) *Ibid.*, paras 37–9.

17) *Ibid., loc. cit.*

18) *Ibid.*, para. 41.

19) Hampson, S. F., *op. cit.*, p. 54.

20) For example, *Inquiry into the Scottish Economy, 1960–61* (Toothill Report), p. 75 and Luttrell, W. F., *Factor Location and Industrial Movement* (London, 1962), I, pp. 319–22.

21) Cameron, G. C. and Reid, G. L., *Scottish Economic Planning and the Attraction of Industry*, University of Glasgow Social and Economic Studies, 6, pp. 23–8.

22) For a discussion of this point, along with evidence, in relation to North-East Scotland, see *North East Scotland: A Survey of its Development Potential* (HMSO, 1969), pp. 103–4.

23) For example, the success of Mid-Wales in attracting small enterprises, especially from the nearby Midlands industrial area.

24) In more ways than one: Gaskin, M., *Freight Rates and Prices in the Islands*, Report to the Highlands and Islands Development Board (1971), especially Ch. 7.

25) There are two striking examples in North-East Scotland. A circle, of eighteen miles radius, around Elgin in Morayshire would enclose about 70,000 people. A similar circle embracing Peterhead and Fraserburgh in Aberdeenshire would contain 50,000.

26) Denton, G., Forsyth, M. and MacLennan, M., *Regional Policies in the EEC* (London, 1968), pp. 298–9.

27) However, with the influx of rig constructors and other oil-related industries into the Invergordon area there is growing anxiety about the labour situation. It is reported that the British Aluminium Company, the original large incomer, has asked the local authority to build more houses to ensure adequate labour for the smelter.

Contributors

Britton, Professor D. K.	M.A., B.Sc. (Econ.), School of Rural Economics and Related Studies, Wye College, Ashford, Kent.
Davey, B. H.	B.Sc. (Agric.), M.Econ., Department of Agricultural Economics, The University, Newcastle-upon-Tyne, NE1 7RU.
Drudy, P. J.	B.A., B.Com., M.Sc., Department of Land Economy, University of Cambridge.
Falk, Sir Roger	O.B.E., Chairman, Central Council for Agricultural and Horticultural Co-operation.
Gaskin, Professor M.	D.F.C., M.A., Department of Political Economy, University of Aberdeen.
Josling, T. E.	B.Sc., M.S., Ph.D., London School of Economics and Political Science, University of London.
Van Lierde, J.	Ph.D., EEC Commission, Brussels.
Ojala, E. M.	D.Phil., Assistant Director-General, Economic and Social Policy Department, Food and Agriculture Organisation of the United Nations, Rome.
Ries, A.	Head of Division, EEC Commission, Brussels.
Rogers, Professor S. J.	B.Sc. (Econ.), Department of Agriculture Economics, The University, Newcastle-upon-Tyne, NE1 7RU.
Schnittker, J. A.	B.S., M.S., Ph.D., Schnittker Associates, Washington DC, USA.
Wallace D. B.	M.A., Department of Land Economy, University of Cambridge.